PREFACE

This text is designed to integrate information in a clear and concise format to allow the general law enforcement and security practitioner to respond to bomb threats, bomb incidents, or chemical-biological-radiological incidents. The text allows the reader to use the information as a general guide and develop local protocols to meet these incidents. With the current threat environment, increased vigilance and knowledge is incumbent upon all law enforcement and security officers whom must have a working knowledge of bombs, explosives and other threats for their own protection. The lack of knowledge can be a contributor to injury, death or the inability to mitigate these incidents. Bombs are far more likely to present a threat, however, the potential for chemical, biological or radiological attacks should not be discounted, and a significant amount of material is presented in this regard. The text allows preparation for such incidents and addresses specialty facilities that require advanced planning such as schools and medical facilities. It allows the practitioner to perform basic threat analysis of potential threats by groups or persons. The common methods to construct improvised explosive devices are covered. An emerging threat in many areas is clandestine drug laboratories. These threats are covered in detail along with techniques to recognize the requisite signatures of a clandestine drug laboratory. This text is not a substitute for trained and well-equipped bomb technicians but does provide the first responder critical information on what protective steps to take and when to seek outside assistance.

J.S.

A LAW ENFORCEMENT AND SECURITY OFFICERS' GUIDE TO RESPONDING TO BOMB THREATS

ABOUT THE AUTHOR

Jim Smith is the Police Chief of the Cottonwood, Alabama Police Department, a former commander of a regional police bomb squad and is an Executive Management graduate of the FBI's Hazardous Devices School. He also holds a Level 2 Certification from the International Society of Explosives Engineers. Smith has performed extensive research, and has published data regarding explosion and fragment mitigation using specialized Class A firefighting foam. He is an active member of the International Association of Bomb Technicians and Investigators and the American Society of Safety Engineers and a graduate of the United States Army's Chemical School Domestic Preparedness program.

Chief Smith is a prolific writer and has published more than fifty articles in refereed public safety journals and several law enforcement and homeland security textbooks. Smith teaches undergraduate and graduate criminal justice classes as an online facilitator for the University of Phoenix and other universities. He has more than 25 years of public safety experience as a law enforcement officer, EMT Paramedic, and is an Advanced Law Enforcement Planner, a Certified Emergency Manager, and is Certified in Homeland Security Level III. Smith has an AS degree in emergency medical technology, a BS in chemistry, and an MS in safety from the University of Southern California.

Second Edition

A LAW ENFORCEMENT AND SECURITY OFFICERS' GUIDE TO RESPONDING TO BOMB THREATS

Providing a Working Knowledge of Bombs,
Preparing for Such Incidents, and Performing
Basic Analysis of Potential Threats

By

CHIEF JIM SMITH, A.S., B.S., M.S.S.

Cottonwood Police Department
Cottonwood, Alabama

CHARLES C THOMAS • PUBLISHER, LTD.
Springfield • Illinois • U.S.A.

Published and Distributed Throughout the World by

CHARLES C THOMAS • PUBLISHER, LTD.
2600 South First Street
Springfield, Illinois 62794-9265

© 2009 by CHARLES C THOMAS • PUBLISHER, LTD.

ISBN 978-0-398-07870-6 (hard)
ISBN 978-0-398-07871-3 (paper)

Library of Congress Catalog Card Number: 2008054991

With THOMAS BOOKS *careful attention is given to all details of manufacturing
and design. It is the Publisher's desire to present books that are satisfactory as to their
physical qualities and artistic possibilities and appropriate for their particular use.*
THOMAS BOOKS *will be true to those laws of quality that assure a good name
and good will.*

Printed in the United States of America
MM-R-3

Library of Congress Cataloging in Publication Data

Smith, Jim, 1954–
    A law enforcement and security officers' guide to responding to bomb
threats : providing a working knowledge of bombs, preparing for such incidents, and performing basic analysis of potential threats / by Jim Smith. –
2nd ed.
        p. cm.
    Includes index.
    ISBN 978-0-398-07870-6 (hard)–ISBN 978-0-398-07871-3 (pbk.)
    1. Bomb threats. 2. Bombs. 3. Explosives. 4. Bomb reconnaissance.
    5. Bombing investigation. I. Title.

HV8079.B62S65 2009
363.17'98–dc22   2008054991

INTRODUCTION

The average law enforcement or security officer will not encounter an actual bomb during their career. However, a probable scenario is officers will have to deal with suspicious packages, bomb threats and found explosives. In the post September 11, 2001 environment, what previously could be ignored or have minimal efforts directed to correct now must be reacted to with significant resources and considered a serious and valid threat. Biological threats have moved from the realm of possibility to a reality. The current evolution of terrorist tactics mandates the full spectrum including chemical events such as the Sarin attacks in Tokyo and radiological incidents are likely to occur within the United States. No facility, location or person is immune from domestic or international terrorism. Those facilities with significant international ties, symbolic value or densely occupied are particularly vulnerable to attack. One must remember the thrust of terrorism has oriented toward soft targets such as schools as in the Beslan attack, theatres as in the recent event in former USSR and other non-traditional targets to facilitate a more deadly and larger scale attacks. This text is written to allow the average public safety official or security officer to deal with these incidents. This material is not a replacement for trained bomb technicians with sophisticated equipment.

Bombings, terrorist attacks and letter bombs have become more commonplace. Statistics published by the Federal Bureau of Investigation report that the actual numbers of bombing attacks have decreased but their lethality has increased. One of the more disturbing factors is the growth of domestic terrorists, including fringe and hate groups which use bombs and arson as an attack modality. The Earth Liberation Front is an example of a group which targets businesses and developments selectively with arson attacks. In the past, bomb threats could be ignored with impunity, however, in the current climate such actions may result in injuries or deaths.

Law enforcement, public safety and security professionals must plan for and respond to incidents that involve bomb threats, chemical-biological threats and suspicious objects. Officers must take the steps as the first arriving personnel which are proper. The failure to recognize and properly react to such an incident may result in catastrophic loss of life.

This text discusses the basic techniques for risk assessment, target and hazard identification. These are essential components in relating to the probability of a bombing attack and the potential outcome of such an attack. The common methods of bomb delivery, bomb construction and methods of triggering are discussed. Letter bombs, vehicle bombs and high-risk facilities such as aircraft, airports, medical facilities and schools are examined. The vulnerabilities and probable methods of attack are discussed.

The utilization of chemical, biological and radiological devices with the unique hazards associated with these devices is addressed. This text includes a section with protocols for emergency medical service personnel in treating those injured from blast, overpressure, shrapnel or chemical agents. Countermeasures and defensive efforts are addressed in sufficient detail to allow the public safety, law enforcement and security professional to make recommendations and to assess any actions taken. Also included are the resources that need to be identified along with a model bomb threat plan. The initial actions to be taken following an explosion and how to protect evidence are also discussed.

This text is not a replacement for trained and well-equipped bomb technicians and is designed to allow the first responder to make identification of suspect items and take appropriate actions until the bomb squad arrives.

The bomb squad should take definitive steps in relation to safe procedures.

ACKNOWLEDGMENTS

This text would not have been possible without the support of my wife, Michelle. The photographs in this text would not have been possible without the assistance of Bomb Technicians and Cpls. John Skipper and Chris Vida. Other bomb squad members who assisted include ABC Agent Brian Harvin, Cpl. Scott Heath, Officer Tim Ellison, Officer Kevin McKee and Officer Tom Flathmann. Also thanks to Bruce Christie of 3DL as a mentor and for his extensive knowledge of bombs, explosives and countermeasures.

CONTENTS

A LAW ENFORCEMENT AND SECURITY OFFICERS' GUIDE TO RESPONDING TO BOMB THREATS

Chapter 1

THE ROLE OF THE FIRST RESPONDER

The first arriving person to most bomb threats, suspicious packages or other suspicious incidents is either a security officer or law enforcement officer. Their role is critical in determining that a threat has been made or exists, its apparent validity, obtaining the needed resources to respond to the perceived threat and notifying supervisory personnel. If the threat is written then evidence preservation is important as significant amounts of trace evidence can be obtained. A verbal threat may be conveyed via telephone, email or in person. Call tracing, and exactly what was said are important. If the threat is associated with a suspicious object or package then the situation needs to be escalated with isolation of the object or package and evacuation of the immediate area. If the threat was an in-person threat then an immediate lookout needs to be broadcast for the person to locate and detain them. Care needs to be taken that the information transmitted contains the nature of the threat such that responding officers are aware of the situation and if a suspect is involved officers are aware they may be dealing with an armed suspect potentially in possession of explosives.

Perhaps the most important factors are to follow agency procedure, perform an initial assessment, obtain the needed resources to begin a response to a threat or suspicious object and notify a supervisor. Upon arrival the supervisor should assume command following the Incident Management protocols specified in the National Incident Management System. The issue of dealing with the management of a facility is always present. Many facilities, particularly government and large-scale commercial operations, will have their own operational plans to deal with bomb threats, suspicious packages and similar incidents.

Coordination and interfacing of resources becomes the issue. In most instances, the security supervisor and/or law enforcement supervisor will be asked for suggestions to mitigate the situation. If the facility is a private concern then usually the ultimate decision regarding evacuation, searches or similar activities resides with their management. Most will want input in the decision-making process. Follow agency procedure and know what resources are available locally or how long they will take to arrive if available only through mutual aid.

Some considerations that must be addressed are:

- Is the threat a ruse to get employees out of the building for attack or to divert resources while another crime occurs such as a robbery at another location or theft at this location? Always consider a sniper attack of employees leaving or outside the building or a bank or other high profile robbery in another location, or even a burglary of the evacuated structure.
- If an evacuation has taken place, did employees look around their work areas before leaving and take personal items with them? This may make a search of some areas unnecessary if employees performed this task. Further, employees would likely be able to identify any items not belonging in their work area.
- If an evacuation occurred have the evacuation assembly areas been checked for suspicious objects, vehicles or persons? Is the area upwind from the suspected device? Preferably the evacuation areas should be scanned for suspicious persons, suspicious objects and the surrounding area scanned for potential assailants prior to evacuation. Always consider the incident a ruse to bring people into an unprotected area for attack with a IEDs or firearms.
- Do the evacuation areas appear reasonably safe in relation to the threatened facility if evacuation is being considered? Make certain no one is standing under window glass as even small amounts of blast overpressure will shatter glass. Falling glass is a serious hazard.
- Is there an explosive detection canine search team readily available? If the threat is credible consider deployment.
- How long will it take the bomb squad to arrive and do they perform searches? Many bomb squads due to the intensive manpower requirements of a search will not perform searches but will

only examine suspicious objects.

- Does the bomb squad or other agency have a chemical sensor capable of detecting explosives through free air "sniff's or swipes of suspicious objects? This may serve as a method to gain information about suspect items or locations as such sensors are specific for explosives and can detect trace quantities.
- What intelligence is available about this location and its occupants? One needs to assess carefully the potential for a credible threat based upon the information available particularly if the location or persons within the location have been involved in a controversy.
- Is this a high-risk facility or occupied by persons who are at risk? Ask about a history of bomb threats or threats made against occupants or the facility.
- Is an evacuation warranted? Possibly, if a credible bomb or other threat exists and management is inclined to do so. An evacuation of at least the immediate area is a necessity if a suspicious object is present.

The most important role of the first responder is to assure that adequate resources are dispatched and a supervisor is informed of the circumstances regarding the call or threat. Coordination with facility security and management is crucial.

Chapter 2

THE ROLE OF THE BOMB SQUAD

Law enforcement and security officers face a variety of unusual situations that mandate innovative and unique solutions. The utilization of resources in these circumstances requires them to have a comprehensive knowledge of the available resources and their capabilities.

A unique resource is an FBI certified bomb squad. These squads have highly trained and well-equipped personnel. Bomb squads may be able to render assistance in non-traditional roles. The bomb squad can be included as part of the multidisciplinary security and safety teams. These teams can assist in their bomb threat pre-plans, which can facilitate their assistance in identification and mitigation of incidents involving bomb threats, chemical-biological-radiological devices or explosives. The bomb squad may also conduct training for those responding to bomb threats. They may have a robot capable of approaching, x-raying and disrupting suspicious packages.

FBI certified bomb squads usually are equipped with two or more instruments that can detect explosive gases, volatile organic compounds and other toxic gases and some will carry devices capable of detecting trace amounts of explosives. Further, many bomb squads will also have screening tools which can screen suspect substances for biological agents, chemical agents and radioactive material. Bomb technicians are trained to operate the instruments and are familiar with operating in toxic environments. Bomb squads typically provide intelligence and threat assessment services for law enforcement and may be familiar with threat groups operating in the area and the modes of operation. These same services are valuable when security personnel are performing threat assessment as part of a hazard or safety program. Some bomb squads have bomb technicians trained extensively in post blast investigation. These investigators can provide valuable technical assistance in the investigation of non-criminal explosions. Bomb technicians are also trained in response and mitigation of weapons of mass destruction from explosives and are familiar with chemical, biological and radiological weapons. The bomb technician can serve as a resource in planning for and responding to incidents of this nature.

Also within the capability of the bomb squads is the ability to perform portable x-rays on objects and some squads have fiber optic observation devices that may be useful. An important factor is that officers are aware of the resources available from their local or regional bomb squad. Some squads will assist with searches following threats while others may not. Some have chemical detection instruments which can detect trace explosives and many have explosive detection canines attached and this may be available to you in credible bomb threat situations. One important factor to be aware of is response time.

Bomb squads that are full-time usually have a protocol regarding their call out and are more likely to be available promptly. Regional or state police bomb squads may have to respond from a long distance. Bomb squads that utilize collateral duty personnel usually are from smaller agencies and may have a longer response time. Military resources are responsible to military or federal facilities first and may assist civilian agencies. The military retains primary responsibility for military ordinance in all settings. If you are using military explosive ordinance disposal personnel as your primary responder due to the downsizing of the military, they may not be as available as in the past. Agencies should establish a relationship with the bomb squad and have a written memorandum of understanding prior to any need. A few agencies will require reimbursement of personnel time and materials for their bomb squad on a mutual aid call. The requesting agency should have a secondary bomb squad available through a written memorandum of understanding should the primary responder be occupied or otherwise unavailable to respond. Remember, *in a crisis* is not the time to be attempting to obtain resources.

Most bomb squads will desire to have a site visit prior to their responding to a particular structure if the structure is classified a high-risk structure. This can usually be worked into a training regimen. Many squads train monthly or more frequently. An onsite visit allows the bomb technicians to familiarize themselves with the potential threat environment the site will present.

Some bomb squads may have explosive detection canines assigned to the squad or attached as part of larger canine unit. Explosive detection canines generally are docile animals with their sole function being the detection of explosives. A few agencies do utilize cross-trained canines that may have another detection function. The explosives that most canines can detect include smokeless powder, black powder, and synthetic black powder substitutes. Additional explosives typically are trinitrotoluene (TNT), ammonium nitrate, pentaerythrite tetranitrate (PETN), cyclotrimethylenetrinitramine (RDX, C4), sodium and potassium chlorate. This provides the canines a very broad range of explosives that can be detected but by no mean is all-inclusive. Many bomb squads now also carry instruments which may through "air sniffs" or paper swipes detect trace levels of explosives. This capability may prove useful in clearing suspicious items or for searches of contained areas.

Deployment is typically in a proactive manner for special events or following a bomb threat. If a large structure or area is to be searched try to use multiple canines to avoid fatigue of the canines and to speed the search. The typical canine will tire after a short period and will have to rest. If others are available, the search can be continued. Canine officers are trained to be spotters and if the canine alerts, he or she will retreat and allow bomb squad members to act. The canine officer usually describes the object and makes a quick drawing for bomb squad members. A method not wise to use is to have the canine inspect suspicious packages previously identified by other methods. This exposes the canine and handler to a potentially explosive device and essentially accomplishes nothing. The suspicious item should never be cleared on a canine's failure to alert. False negatives are a strong possibility in explosive devices. The canine is a screening tool only. If the item is suspicious, the bomb squad should handle it. The same is true regarding instruments indicating trace explosives or not indicating trace explosives. Such should not be used as the sole method of clearing a suspicious item.

Chapter 3

TYPES OF BOMBERS

There are a variety of bombers that range from amateur to professional with possibly more than one motivational factor. The amateur or experimenter is described as making devices that are crude and unsophisticated. These devices typically are delivered against targets with low security or may be in an open field or vacant lot. They may have sophisticated timing devices but use small amounts of main charge since their primary purpose is for excitement, not destruction. Smokeless and black powder bombs are typical since they are readily available. This may be a thrill-seeking teenager or someone at any age seeking excitement. Some of these devices are made from chemicals that derive an explosion from overpressure such as mixing isopropyl alcohol with high test chlorine (HTH) powder in a container. The resulting overpressure explosion can be impressive.

The professional may be a mercenary, terrorist or an operative in a crime syndicate who produces high quality devices or uses sophisticated operational techniques. The device placement is to assure major damage or death. Time and study is used in making and placing the device. Significant intelligence is obtained regarding the target. These devices may be part of a series bombing. These devices are improvised but in many instances their level of sophistication may be high with remote command radio frequency devices causing detonation ranging from garage door openers, cellular telephone, portable radio transceivers or pagers. Eric Rudolph used a remote command detonation device in the Birmingham Alabama abortion clinic bombing and killed the police officer attempting to examine the suspicious object concealing the bomb Rudolph had earlier emplaced near the clinic entrance. The common tactic used to engage troops in the Iraq and Afghanistan theatres is a remote command detonated device using an IED from unexploded military ordinance. The devices are detonated as soldiers approach the device. Other terrorist will use command detonation devices to detonate hand carried, concealed within a vehicle or concealed explosives upon their person in the suicide bomber scenario. These devices are usually simple toggle switches with a power source to a blasting cap to detonate the main charge.

The psychopathic bomber acts without apparent rhyme or reason with little or no predictability. The construction of the device is not usually predictable; however, there may be signature evidence that links the bombings. Typically this is a "serial bomber." Examples include the "Mad Bomber" George Matesky and the "UNA-BOMBER."

The motivation of bombers varies and may be for a single or multiple reasons. Some of the common motivations are listed below:

- Ideological–The bomber is following a belief or ideology; usually extreme left or right wing. These bombers tend to build more professional devices. They may have radical political or radical religious beliefs. Racial or ethnic hatred is not uncommon. Many terrorist bombers would fall within this group.
- Experimentation–Typically this category consists of youthful or immature offenders, who are drawn by the excitement and noise. They may stay and watch the device explode or to see if it actually works. This category usually consists of amateurs and many are juveniles. These individuals in many instances will construct incendiary devices and use readily available materials. Some may use bottle bombs made from corrosive material such as drain cleaner and aluminum foil or gunpowder obtained from disassembly of bullets. Common hobby cannon or firework fuses are used as the delay in the firing train. Many use soft drink bottles or cans as a container with the rare metallic or pipe bomb constructed by this group. Very rarely one may find this group manufacturing more sophisticated explosives such as ammonium nitrate mixed with nitro methane or TATP.
- Vandalism–This category commits destruction for the sake of act itself. Many are amateurs, may be experimenters and are usually juveniles. One common finding is the use of bottle bombs and small devices to destroy mailboxes.

- Profit–Bombings in this category are used for direct or indirect profit. This may be an organized crime operation that finds bombings or arson useful in extortion and destroying a competitor's business. Some individuals will use bombings for insurance fraud and rarely bombings are used to cover up another crime. Professionals usually populate this category; however, some "first time" bombers may be found trying to commit insurance fraud. The devices used are normally sophisticated and incendiary devices are common within this practice.
- Emotional Release–These bombers are usually psychopathic bombers who are seeking release of real or imagined transgressions by the target. Many may have sexual connotations related to the bombing itself.
- Revenge–This group is closely associated with the emotional release bombers. They may be motivated by real or imagined transgressions by the target with some being domestic related. The domestic bombings usually are well constructed devices and are victim initiated.
- Recognition–This individual creates devices to become a "hero". Infrequently these bombers are found among public safety personnel who attempt to increase their stature by being first on the scene. Firefighters starting fires with incendiary devices are one of the more common violators.

Targets of bombing attacks include a variety of factors:

1. Anti-personnel attacks may be used to generate large numbers of deaths or injuries. This is a common terrorist tactic whereby large number of persons are injured and killed. A variation is a letter or package bomb directed at a specific target or targets. A series bombing may be used to successfully attack several locations of the same business. This tactic is both a terrorist tactic and one commonly used by individuals who seek revenge for a perceived wrong.

2. Symbolic bomb attacks are carried out against government buildings, corporate structures, or military facilities. A release to the media many times follows from the group or person taking credit for the bombing. These may be signature attacks that follow a specific pattern or specific set of targets. Terrorist groups such as the Earth Liberation Front use this tactic to attack residential developments and sport utili-

ty vehicle dealerships with incendiary devices.

3. Selected target attacks are against a particular set of targets with a common theme. Examples include multiple letter bomb attacks against selected diplomats, executives, animal experimentation facilities, government structures and related targets. This tactic is used by terrorist groups and was a specialty of the IRA.

4. Prolonged bomb attacks are designed to draw attention to a particular cause and may used to gain release of prisoners. The Basque ETA and IRA both have used this tactic as did the Mad Bomber.

Targets of bombings may be a particular target for a specific purpose or the target may be an indiscriminant target of opportunity. Attacks may be to acquire assets, as in bank or armored car robberies that use bombings to distract police, or to induce terror in the population, or attract media attention. An aborted bombing led to one of the more dramatic police shootouts in U.S. history in early 1980s in Norco California. The main charge on a natural gas line failed but did not stop the bank robbery. During the subsequent shootout with police the robbery suspects fired more than 1,000 rounds from assault type weapons, destroyed more than 30 police cars, shot down a police helicopter and used numerous homemade grenades both of the hand delivered and shotgun launched type. More than a dozen officers were injured by gunfire and explosions and one officer was killed by gunfire.

The attack type preference for many groups is: bombings, arson, assassination or mass shootings, hostage taking. Attacks by bombs are usually directed at times of peak occupancy to produce numerous casualties and more media attention.

Bombings are very effective tools. Rarely are bomb warnings provided. Terrorist and criminal apprehension probability in bombings or arson is very low. Extortion is occasionally used. Sabotage is becoming a more common event particularly when it affects a technology such as power lines, nuclear power plants, highways, trains, or pipelines.

Bombs also provide an inexpensive impersonal method to attack a target. The bomber can be a long distance away in physical distance and time when the attack occurs. The risk of being caught remains low as is demonstrated by long lived attacks conducted by the Mad Bomber and UNABOMBER.

BOMBER THREAT ASSESSMENT

The following information is derived from a variety of sources, which include published information from the U.S. Secret Service, Bureau of Alcohol, Tobacco, Firearms and Explosives, and National Institute of Justice. The information presumes that the suspect in bombing attacks fits typical behavior models as many assassins utilize explosives to accomplish their goal of assassination. The motivation of bombers is offered to expand the knowledge of bombing suspect behavior and make identification of suspects easier.

Bombing attacks whose purpose is the death of a particular person or persons, usually have motives that are presumed to be other than domestic, drug or gang related. Many of these motivations will be political or ideological. The targets may include elected officials, celebrities, judges, business executives, state and local officials. The attack may be used to gain personal attention of the attacker such as the John Hinckley attack on President Reagan where Hinckley tried to gain Jodie Foster's attention.

The literature emphasizes that no true "assassin profile" exists. Many do have common features such as mobility and transience. Most are socially isolated and have a history of weapons or explosive use or at least ready access to such. The commonalities of school bombers are discussed later in the text in the section relating to school environments. They may have been little law enforcement contact but usually the suspects will have a history of harassing other persons. Poor impulse control with explosive angry behavior is normally exhibited. They even may be an indicated interest in attacking a public figure. The bomber may have an interest in radical and/or militant ideologies but is not usually a group member. Many will have a history of depression and attempts at suicide. Almost all will have a history of a grievance or resentment against a public figure or an institution whether real or imagined. George Matesky, the "Mad Bomber" had a perceived grievance with Consolidated Edison resulting in his multi-year bomb attacks. What is odd is Matesky relented during World War II for patriotic reasons but continued his bombing attacks following the closing of hostilities. The UNABOMBER was concerned with technology ruining civilization and conducted his campaign to make his position known demanding his manifesto be published. His manifesto was recognized when published and ultimately led to his arrest.

The literature also emphasizes that most of these individuals do not have a significant mental illness. Mental illness usually does not play a large role in their bombing attack. The behavior is not deranged behavior in sense of legal insanity. Many have emotional and mental problems, which may contribute to the bombing attack, but usually the mental illness is not the root of the problem. There appears to be no "snapping" of the person as many attacks are well planned and methodical. However, a precipitating event such as a personal loss or personal failure, real or imagined, which starts the bomber on the last phases of preparation and launches the bombings.

Many persons who carried out bombing attacks did not make any direct threats. However, some did provide indirect threats or indicate a grievance to others. Persons who are dangerous may not have made a direct threat. However, they may have provided indications of their intent in several different manners. These individuals usually have made verbal or written statements which indicate they are considering a bombing attack. The most common recipients of this information are family, friends or coworkers. The persons making direct threats should also be carefully evaluated for the threat they pose.

Bombers usually utilize the bombing attack to attain a goal they wish to achieve, even when the motive may be unclear to other persons. The motives and selection of the target are directly related. These acts are usually not impulsive or spontaneous. Some may be well thought-out including building test devices and rehearsals which are also a common finding in terrorist bombings. Bombing attacks are usually triggered by some seminal event in the bomber's life that makes them seek to utilize violence against a discrete target as a problem solving event. The bomber is usually under or has tolerated some dramatic stress and seeks violence as an outlet. Other outlets for stress include: physical illness, psychotic behavior, suicidal behavior, and violence. The event may be a variety of situations but the most common stressors are loss of a significant other, financial distress, and changes in living status, being rejected or humiliated; or failure in some significant portion of life. The event does not even have to be real, only perceived to be real by the individual.

In many instances, the motive for a bombing attack may be for notoriety; to bring attention to a problem or revenge for a perceived wrong. The motive may be to save an ideal, to make money, or to have a special relationship with the victim. Many bombings are committed to attempt political change. The major categories include facilitating attention to a particular problem or attempting to make political change. Gaining notoriety or fame follows this. In many instances, the bombing target was not selected due to hostility toward that individual or institution but based upon motive, target availability, and ease of bombing attack.

One of the key elements of any threat assessment program is to identify the potential bombers, their assessment, and tracking. Many persons self identify by making open threats verbally or in writing. Those persons with an unusual or inordinate interest in the protected figure or facility should be identified and assessed. All personnel who are the staff of any official should be trained to identify persons of this nature who exhibit an unusual interest in the facility or protectee.

Assessment of the person should, as a minimum, include: local police arrest and records check; national criminal history; public records credit history; drivers license, and motor vehicle records check to identify their vehicles; obtain a photograph which is usually available from their driver's license or state identification card. Any military or federal service, travel out of country along with their current residence should be identified. The person's education, current employment, local and distant relatives should be identified. Any previous mental health or emotional problems along with their current health is important. Prior experience with bombs, explosives, military training including college or other chemical or electronics knowledge is also important. If parents or siblings are available and cooperative, they should be questioned as to the suspect's predilection as a juvenile for "playing with fire," arson, inappropriate "bed-wetting," or cruelty to animals. These behaviors may be the manifestation of poor impulse control. Many investigators believe that when these factors are present or if any single factor is present, will make the suspect a good candidate for homicidal behavior. The interview of neighbors, employer, coworkers, relatives and friends usually provides much of this data. As much of the data obtained should be corroborated from more than one source when feasible to assure validity and reliability and is true regarding threat statements or ideology of the suspect.

Usually many of these factors when taken in isolation do not indicate that the suspect is considering a bombing attack on the target. However, upon investigation there may be significant evidence that the suspect is planning an attack. All threats, inappropriate attention, unusual requests for information must be carefully evaluated. Each factor listed above must be considered. This may indicate that the suspect is making an overt attempt to attack through use of a bomb.

Direct interview may be an option. The minimum information that should be obtained from the suspect includes: demographic information, education, military history, criminal history, history of violent behavior, medical history, marital relationships, mental health history, grievances, history of harassing others, interest in radical ideology or groups, travel history, current life situation and circumstances. The suspect should be directly questioned about their ability to assemble and utilize bombs. The suspect should also be questioned as to their ideas or plans to attack a public figure. If the suspect has indicated or had an unusual interest in a public figure, they should be asked about that interest. Another area of particular interest is whether the suspect has stalked or visited the target's work, home or other frequented sites for reconnaissance purposes.

Another important factor to explore what, if any, motives or grievances the suspect has. The bombing targets of such individuals may change several times during the episode prior to attack. Perhaps the most dangerous and difficult circumstance to mitigate is the suicidal bomber. This individual may be ready to sacrifice their life to bring attention to a particular issue or right a perceived wrong. Questions in regard to suicidal ideation are important in this regard. A personally delivered "human bomb" is very difficult to stop. The average person may be able to carry up to 45 pounds of explosives on their body, or hand carry a container. When such charges have shrapnel added their killing range is impressive and may be up to 300 feet or more in open areas against a lightly armored target. The driving of a "suicide bomb" vehicle into a building or in proximity to the target may allow a suspect to deliver up to several tons of explosives leading to widespread destruction over a several thousand foot radius. The key to any comprehensive bomb protection program is the gathering, evaluation, and follow-up on intelligence. All key personnel must be trained to recognize individuals who may pose a threat to an official or facility.

Chapter 4

RISK ASSESSMENT

There exist two specific target identification and risk assessment methods, general and specific. Most assessments are subjective. High value targets should receive the benefit of the doubt and receive a full response. The general method compares the facility and its occupants to target profiles looking for items that would suggest their availability as a target. Items to consider are:

- Is the target open to attack with no substantial protective measures in place? Consider what protection is in place or is reasonable to put in place.
- Is the target suitable for attack? What groups or individuals are interested in this facility or might have as interest in the target? Multinational companies may have domestic and international attention.
- How vulnerable is the target? Think about what could be done to attack the target with mail bombs, letter bombs, hand placed devices or a vehicle bomb.
- Does the target have symbolic value and importance to the general public? Once again, what groups or individuals have an interest in the facility?
- Are the occupants a target particularly if in a densely occupied area with an attack likely to produce mass casualties? Are there a large number of persons housed within the facility or location?
- How essential is the target and how easily restored if successfully attacked? What sort of impact would loss of the facility impose on the community and business? The loss of a power generating plant or communications complex might have substantial impact to the region.

Can risks be quantified? The answer is yes but remember the process is subjective and second-guesses a potential adversary. One technique that uses numerical values which are assigned to risks with a yes value of one and a no value of zero. Scores of one to three are moderate risks with scores of four or more considered a high risk. An adversary such as a mentally disturbed person or fringe group may not have a rational or logical reason that is apparent for attacking a particular target. This is true if the grievance is of a personal nature against one individual as is common in domestic disputes.

Examples of targets that might be easy to attack are pipelines, storage tanks, fixed infrastructure such as bridges or power lines, but are they suitable targets to attack? If the purpose is disrupting the infrastructure or if the group or individual has a grievance with the owner, these targets may be suitable. In fact, the attack on a pipeline may not be what it appears. The attack might be slated to damage the land the pipeline crosses and leaks product if an individual or group had a grievance with the land's owner. Or the attack may be simply one of vandalism. The sophistication of the attack my lend credence to the motive as normal vandalism attacks do not use sophisticated or large devices.

How vulnerable is a target to attack and what protective measures are in place is a factor that must be assessed. Easily attacked targets are inviting therefore any protective measures such as basic security, perimeter protection, package and vehicle screening may be a deterrent to many potential attackers. Simply because a target is vulnerable may not in and of itself make it a suitable target. Many groups and individuals will be attempting to make a "statement" with the attack such that a suitable target must be easily recognized and receive media attention. Usually to garner media attention multiple injuries or deaths are needed. Another alternative is the target itself may not be recognized but its disruption may attract media attention and the public's attention. The disruption of infrastructure such as power lines, waterworks, and communication facilities would meet these criteria. The loss of basic utilities would also impact the public in making them feel they no longer control their environment. While an inconvenience, the majority of the public would not be dramatically affected. An attack on waterworks with the loss of water supply would have a more dramatic impact and would impair the delivery of fire services. Arson attacks coupled with loss of water supply could be catastrophic. In this manner a coordinated attack that individually might not produce panic or significant disruption could become disastrous. Another feature of an attack may be to induce terror and panic. An attack upon a chemical production facility, hazardous materials storage facility or nuclear power plant with the potential for release of toxic hazardous materials might induce panic.

This data provides a general idea of the target factors but does not identify the specific targets of attack. An individual may have a grievance real or imagined with a person or entity that meets few if any of the criteria. Do not discount the potential target may not be the facility or physical structure but one or many of its occupants. This is particularly true of domestic-related bombings. If a potential target reports a domestic disturbance and has received a suspicious item or threat use caution. Domestic bombings normally involve IEDs which are victim activated. The candy or flowers delivered may conceal a bomb.

What about specific targets within a structure? The same risk assessment can be applied. The obvious points to attack in any structure are those that inflict injury to the occupants or damage to the structure. If the point is to make the structure uninhabitable an attack on the high volume air conditioning system to disable the system is one scenario. That same scenario could be modified to use the system to deliver a chemical or biological agent to the occupants. This has been accomplished in several locations when oleoresin capsicum (OC) better known as pepper spray or traditional tear gas has been introduced into ventilation returns. The agents were then distributed throughout the facility and effectively denied the use of the facility. Schools, businesses and shopping malls have been attacked in this manner usually as a prank. These agents are normally not lethal but the ease with which they were distributed in a densely occupied structure such as a shopping mall is disturbing. Other points of attack on a structure for denial of use could include power lines, natural gas lines, communications equipment or other essential infrastructure. However, the more common form of attack would be to injure or kill the occupants of a structure.

Chapter 5

BOMB THREATS

The typical bomb threat is a short blurted sentence on a telephone stating "There is a bomb in the building!" What are the steps to be taken when a threat is received? The law enforcement and security officers will have to make numerous decisions that may or may not be covered by written procedures. Management of the facility may range from lackadaisical or hysterical regarding the threat. All bomb threats should be taken seriously.

The most important aspect is to determine exactly what was said if the bomb threat was via telephone. The more information provided tends to make the threat more credible. What is the background of the target of the threat? Was a specific person named? Is the threat aimed at an individual or a company? Is there any labor, domestic or other controversies involving the person or company? Has a suspect package been received or located?

A bomb threat credibility assessment sheet follows. This information should be obtained and the sheet scored. Scores of more than three indicate a credible threat and additional actions such as evacuation and search should be considered.

LAW ENFORCEMENT AND SECURITY THREAT ASSESSMENT CRITERIA

Time and Date Call Received:_____ Caller ID or *57
Trace Used? Yes No

Exact Words of the threat:_____

_____

Threat Assessment: (Verbal and Written Threats)

Did the caller say or does the note read:

Where is the bomb?	Yes	No
When will it explode?	Yes	No
What does it look like?	Yes	No
What type bomb is it?	Yes	No
What can be done to stop it?	Yes	No
Was money demanded?	Yes	No
Was anything demanded?	Yes	No
Was Trace or Caller ID successful?	Yes	No
If successful is the party known?	Yes	No
Is the threat aimed at a specific person?	Yes	No

Score Yes 1 Point Each Total Score_____ *Score >3 Are Credible Threats*

Target Assessment Factors:

Prior Bomb Threat(s) in last year?	Yes	No
Employee-employer controversy?	Yes	No
A specific person named?	Yes	No
Specific person has domestic problems?	Yes	No
Labor dispute?	Yes	No
Government Facility?	Yes	No
Government Official named?	Yes	No
School?	Yes	No
Medical Facility?	Yes	No
Recently terminated employee	Yes	No

High Risk Facility?   Yes   No
(Jail, police facility, courts, lawyer, judge, abortion clinic, hospital, physician, animals at facility, timber business, environmental related, media, international business).

Recent news story on facility or person?   Yes   No

Score Yes 1 Point Each Total Score_____   *Score >3 Are Credible Threats*

The threat assessment focuses upon information provided in the bomb threat. The more information provided, the more substantial the probability the threat is a credible threat especially if the device is described in detail. Some bombers are proud of his or her devices and may provide details concerning its construction. Extortion coupled with a bomb threat is common, particularly if money is demanded. This is a common bank robbery method and is sometimes accompanied by a hoax bomb. Other demands may be made in extortion threats. If call trace is successful, does security or police know the person to whom the telephone is listed? This is an important piece of information. Client and employee lists may provide a connection to the caller. If the threat is aimed at a specific person, the circumstances surrounding this person should be examined closely. In some instances, a top management figure or other visible person may be named in the threat. Federal Judge Vance killed in Birmingham Alabama by a package bomb was victim of a bomber whom Judge Vance had ruled against in court. The individual had allegedly written anonymous threat letters and expressed his anger to his family members regarding the court ruling by Judge Vance prior to the bombing.

Target assessment factors review the visibility of the facility or person threatened and examines any factors that may result in the facility or person being a target. Examples of factors that lend credibility to a bomb threat would be an existing controversy such as a labor dispute, publicity about the facility or the person threatened, or a recently terminated employee under acrimonious circumstances. Obviously, some facilities or persons have a greater risk of receiving credible threats such as schools, government facilities and officials, public safety facilities and the like. These facilities are typically heavily occupied and present a significant target and are rated higher.

If the bomb threat was made by telephone, take steps to trace the telephone call. Call the telephone company security and request a trace on the telephone line the call was received. Officers should have a relationship with telephone security personnel and know what capabilities exist within the telephone company. If the call originated from inside the structure know the capabilities of the particular telephone switch in use. Determine if the telephone has calling number identification or other security features in place and if the call was recorded. Many telephone companies offer a call trace feature activated by the keying of *57. This feature allows the telephone company to promptly trace the call location. If the call was recorded, make certain the recording is retained for evidentiary purposes. Review the tape if possible. If a telephone number is obtained, determine to whom the telephone is listed and the location. Establish if this individual has any connection with the target. Another law enforcement or security officer should be sent to the location of the telephone to conduct an interview. In many cases, telephoned bomb threats will be traced to pay telephones in a public location or to a cellular telephone. It may be possible to lift latent fingerprints from the telephone that allow identification of the suspect using automated fingerprint identification system. This may be particularly important in serial bomb threat cases. Some jurisdictions classify telephoned bomb threats as misdemeanors while others classify such threats as a felony or by the nature of the target such that hospitals, nursing homes, schools and similar facilities are felonies. Know your jurisdiction's law regarding telephonic, in person, and written bomb threats. Threats to a facility engaged in interstate commerce or receiving federal funds may be a violation of federal law. Threats made to utilize a chemical, radiological or biological weapon are federal violations and fall under the jurisdiction of the Federal Bureau of Investigation.

Do not use the telephone the call was received on to make any calls and if possible leave it "off hook." This may not be possible or feasible in the cases of switchboards or multi-line telephones. Even if security features are not in place, some telephone companies can trace the last outbound or inbound call on a particular telephone line. Have the person receiving the call write down exactly what was said and the time the call was received. Have them attempt to remember any distinct voice characteristics or background noises that might have been present. This may be of value in serial bomb threat cases, particularly if the same location is used to make the threats.

Data from the Federal Bureau of Investigation's Bomb Data Center indicates few actual bombings are preceded by a threat. However, this does not mean that the bomb threat should be ignored. The rationale behind most threats is to have the bomb threat serve as a nuisance or to damage the target by disrupting operations. Some individuals hope that the negative publicity surrounding the bomb threat will be reported by the media. For this reason every effort should be made to prevent knowledge of the threat from becoming public or reported by the media. Bomb threats are disruptive particularly when activities are in progress that is sensitive to disruption. A common bomb threat for schools is to receive bomb threats during testing of students. Occasionally, a disgruntled employee will make a bomb threat to "shut down" a business while an evacuation and search are undertaken. Always note in the investigation of the telephoned bomb threat that confirmation is made that the threat originated from an outside telephone. Threats made on interior telephones narrow the scope of the suspects but make detection of a device less likely since employees make have unhindered access to many sensitive areas within the facility.

The form with the questions to ask should be placed near the telephone switchboard or other locations where a bomb threat might be received. If a caller is cooperative they will provide the information requested, however, this is unusual as most callers provide the threat then hang up the telephone to minimize their exposure. Usually the more information that is provided, the more credible the bomb threat becomes. This is true especially if the threat provides a detailed description of the bomb itself.

BOMB THREAT SHEET

Time the Call was received:_____

Where is the bomb?_____

When will it explode?_____

What does the bomb look like?_____

What can be done to stop the explosion?_____

Who is this? (This usually prompts a hang up)_____

Write down exactly what was said. _____

Questions for the responding officer to ask:

1. Was Calling Number ID, *57 Call Trace, or other security feature used? Was this an outside call? Was the call recorded?
2. Have you contacted the Telephone Company? Has the telephone been used since the threat was received? Was a trace successful?
3. Have you received any unusual mail, package or other items?

Ask management:

1. Are there any controversies affecting the location?
2. Have there been any employees recently terminated or disciplined?
3. Does management want to search or evacuate?

WRITTEN AND IN PERSON BOMB THREATS

One consideration regarding the written bomb threats is that the envelope may be the vehicle for a more sinister threat. It has become popular to place various powdery substances within the envelope and allege the substance is anthrax spores, Ricin, a dangerous chemical, the highly unstable and sensitive explosive TATP or radioactive material. Exposure to anthrax spores or Ricin is potentially fatal. The vast majority of these incidents are hoaxes according to the Federal Bureau of Investigation. However, specific steps should be taken which are addressed in the chapter on Chemical, Biological and Radiological Devices. If this occurs it escalates the simple written threat to a full-blown incident that may require substantial outside resources to contain the potentially harmful substance, decontaminate those exposed and to confirm the substance is, in fact, harmless.

Occasionally an individual may make the threat in person. This usually allows at least a description if not identification of the suspect. Any items touched or handled by the suspect should be isolated and retained for identification if the suspect is unknown. These threats should be handled in the same manner as a telephoned threat. You should be familiar with your agency's policy and the local laws regarding in-person bomb threats. The policy should state the actions to be taken if a suspect is identified as to whether you can make a probable

cause arrest or whether a warrant will be required. If the suspect has
been detained or present you should determine if you should interro-
gate the suspect or call for an investigator. Be familiar with your
agency's policies. A crucial consideration is the person armed or concealing a bomb? If the potential for this scenario exists one should
have a written policy as to the manner the person is approached and
detained. A suicide bomber must be considered and the police deten-
tion of a person making threat versus a person whom is a suicide
bomber is radically different. Some agencies authorize deadly force to
detain a suspected suicide bomber.

Suicide bombers have become a reality. These individuals load their
body under clothing with explosives and walk into the target area and
manually explode the device although some suicide bombers may
have a "minder" with a radio frequency back up command detonation
device. Some may use vehicles loaded with explosives or other means
to deliver their deadly cargo.

Another consideration is whether the person has on their person a
bomb, firearm, or other weapon. Specific officer safety steps should be
taken to confirm whether the individual is armed. The suspect should
be detained, handcuffed with his or her hands behind their back so he
or she cannot potentially access a switch for a device worn on their
person or in a hand-carried item. Metallic handcuffs should be applied
over clothing to prevent conduction of electricity. Some suicide
bombers have contacts separated with one in each hand or arm such
that clasping the arms or hands together may complete the circuit and
detonate the device. If during a weapons search any wires or other
objects are found, attempt to further immobilize the person. If possible, move the person to an area away from the structure and people.
As a minimum, evacuate the area and call the bomb squad.

These devices are, by their very nature, limited in size and com-
plexity. This is a preferred method of terrorist bombing in some areas
since no rescue or escape plan is needed. The suicide bomber will be
dressed to blend in with the environment when possible but some do
not fit such as a person wearing a long-sleeved shirt or coat for concealment of explosives and wiring in warm weather. This individual
may not fit what is perceived as the normal appearance of a suicide
bomber. These devices normally use high explosives augmented with
shrapnel in the form of metal sheets, nails, glass marbles or bolts with
electric blasting caps. The detonation switch is usually a simple toggle

type or push-button switch that can be operated under duress. The explosives may be worn distributed around the torso covered by clothing, may be worn only on a portion of the body, may be concealed in clothing or may be carried in a handheld bag, box or piece of luggage. Although limited in the amounts of explosive that can be concealed in or under clothing, bombers can conceal several pounds of high explosives. This amount of explosive is sufficient to kill persons in a large area particularly if densely occupied and if the explosive is shrapnel enhanced.

In some instances, the individuals will wear a vest over clothing to conceal or contain the explosive or wear loose or baggy clothing to conceal the explosives. Scan suspects for this type of clothing or the appearance some objects are concealed within or under the clothing. Remember, innocent-appearing hand carried appliances such as a radio or other electronics may conceal explosive devices. Backpacks or book bags can easily be used for this purpose.

Always consider a suicide bomber may have already placed other explosives. Search the areas the potential bomber had access to for secondary devices. Remember, vehicles operated by a suicide bomber may contain explosives and there may be additional accomplices or suicide bombers present.

Chapter 6

SEARCHES

If management elects to conduct a search of the facility, several steps should be taken. Persons familiar with the areas should perform the search. If law enforcement or security officers are to search, a person familiar with the area should accompany them. Searches within a facility by persons not familiar with the facility are inefficient and ineffective. Some bomb squads will assist with searches while others will only respond if a suspicious item is found. Know what your local bomb squad's policy is regarding searches. The search should be for any item that cannot be identified as belonging in a certain area by the persons who work or are responsible for the area. The search should be minimally intrusive and not disturb any suspect items. Searches should be conducted prior to evacuation by those persons whom work in the area and may consist of employees simply looking around their work areas for something that they do not recognize or did not bring with them. Remember, a search may trigger a device.

Search public access areas first. Examples include lobbies, hallways, restroom facilities, waiting areas and the like. The areas most accessible to the public particularly that are in proximity to entrances or exits are a likely location. Attention should be paid to unattended bags, boxes, or briefcases. Other likely areas are lobbies, auditoriums and public restrooms. Particularly difficult to search are areas with suspended ceilings with panels that are easily removed or pushed up to secrete an improvised explosive device (IED). This facilitates concealing a device into the overhead. Other items that require attention are trash cans, ashtrays and flowerpots. Decorative plants or shrubs in a flower bed or planter can be used to conceal a device. A casual bomber will usually utilize the most easily accessible and least observable location to deliver a device.

Any incoming mail and packages should be carefully screened. The key to detecting a suspect package is one that is not expected. If a package is located following a bomb threat that is not expected it should be considered suspect. The sender should be contacted to see if the package is a legitimate package. The chapter on Letter Bombs discusses other indicators of letter and package bombs.

ROOM SEARCHES

If a detailed search is undertaken, the caveat to remember is the search itself may trigger a device. The proper way to search a room is with two searchers. The searchers should be familiar with the room. The search starts with the searchers locating themselves in the center of the room back to back. They remain there quietly for several seconds with their eyes closed to listen to the ambient noise. Any unusual noises are investigated initially. The search starts at desk level with the searchers dividing the room in half proceeding in opposite directions. The next level searched is at the floor level. Care is used in opening closed cabinets or drawers. The final sweep is at above-desk level and ceiling level. Suspect items are not disturbed but noted and the search stopped. Suspect items must be examined and cleared by a bomb technician. Schools present a special problem with many hiding places such as lockers. Book bags and backpacks are also a common item located in schools. It usually is not feasible to identify the owner of these items. Review the chapter on Schools for additional information.

When it has been decided that a search will be made, it should be made on a systematic basis. Each area checked should be marked with colored tape or other markers to indicate that the area has been searched and efforts will not be duplicated.

Next, divide the room in half and begin the search. The search is done in three separate sweeps. The first sweep is from the waist down and the second is from waist to eye level. The third sweep is everything above eye level. Do not forget suspended ceilings.

When searching, do not be afraid to be nosey. Check everything, but if you find something, do not touch or move it. Report any suspicious items to a supervisor and follow agency protocol. Some agencies will use a thermal imager to assist in searches hoping the bomb may generate heat from a battery powered timer generating heat or have a burning delay which might be detectable.

VEHICLE SEARCHES

Non-bomb technicians may undertake routine examination of vehicles. However, if a bomb threat or suspicious circumstance exists surrounding a vehicle, bomb technicians should examine the vehicle. The technique listed below is typically followed by bomb technicians and may be used as routine examination of vehicles for security purposes.

Scan the area first and then search the surroundings. If feasible, visually examine the vehicle from a distance (binoculars). Some agencies also use a thermal imager looking for temperature anomalies which might indicate an IED. Check trash cans, dumpsters, boxes, planters, objects and other vehicles near the suspect vehicle. Remember, the device may be on a timer, using movement or anti-tamper initiation for detonation. Avoid tunnel vision. Do not touch the vehicle. Do a walk around, look externally for any unusual features, wires, trip wires, visible devices, and pressure devices including under the tires, and consider booby traps. Look inside the vehicle from the outside without touching and look for unusual features.

Use a mirror to examine the underside of the vehicle starting with the engine compartment. Do not touch the vehicle. Work from front to rear along both sides searching the left side first. Areas of interest are: under the driver's seat area, top of the muffler and exhaust pipe, in wheel wells, gasoline tank, catalytic converter, wires attached to spark plugs or distributor, and against the firewall on drivers side. These are all common locations for bombs. Use the mirror without touching the vehicle. It may be necessary to slide under the vehicle to examine all areas.

Carefully examine all areas accessible near the hood and hood latch. Open the hood remotely. Carefully examine for wires, trip wires and pressure relief devices by mirror before disengaging the second latch. Disengage the second latch remotely if possible. Repeat the process and use the mirror when possible. Carefully move the hood up enough to use the mirror over the engine area to search it. It may be necessary to prop the hood open. Use a plastic or wooden prop, not a metal prop. Carefully examine all areas of the engine compartment. This is common location for bombs.

Determine the location of light switches and, if possible, look at a vehicle of the same year and model prior to touching or approaching the vehicle. A wiring diagram if available (a rarity) is excellent to determine the potential location of devices and booby traps. Carefully examine the right rear passenger door. Open it remotely. Examine the interior with the mirror paying attention to the driver's seat, under the seat, under the dash, and all accessible areas. Feel the surface of the seats and then open the other doors remotely if possible, or alternately from the inside. Open the glove box remotely and examine it. Look at all times for wires, trip wires, pressure relief devices, or similar indicators. Examine the rear seat by removing and rear areas of the vehicle. Under the driver's seat is a common location for bombs.

Open the trunk remotely if possible. Examine the trunk thoroughly. Check the spare tire. Pull up any fabric coverings and check the "wells" on both sides of the trunk. Start the vehicle remotely if possible and let it idle. Operate all switches and move the vehicle through all gears and move the vehicle. If a device is found, immediately evacuate the area. Do not forget *time, distance* and *shielding.* If you can see the device, it can hurt you.

BOMB INCIDENT MANAGEMENT FOR SPECIAL EVENTS

The potential for a bomb threat or a bombing to disrupt a densely populated event such as the fair, sports events, parade, or other related events, are incidents that must be planned and mitigated. The scenario of most attacks of this nature will be a hand delivered device although one cannot rule out a large vehicle borne improvised explosive device (LVBIED). Security and law enforcement personnel working at any event or venue should have had entry-level bomb incident management training as part of the overall emergency plan and training.

Law enforcement officers and security personnel should inspect parade routes, VIP seating, and other venues prior to the event start and/or before occupancy. Suspect items identified will be brought to the attention of the supervisor and if needed the bomb squad requested.

A hand delivered device will most often be placed in a public access area and will be disguised. Security and law enforcement personnel must be constantly looking for items, boxes, packages, purses, backpacks or briefcases that are abandoned, or in a high-risk location. Upon locating such an item the security officer or law enforcement officer should alert a supervisor and attempt to locate the owner. The supervisor will notify the bomb squad if an owner cannot be located promptly. Nearby persons and employees will be questioned to determine if they observed anyone leaving the item. If an owner or other reasonable explanation for the item cannot be determined within 15 minutes, the item should be declared suspect and a local evacuation will be made (no less than 300 feet but preferably 1,500 feet). Suspect items will not be disturbed, inspected or moved. The bomb squad should be requested. The on-scene supervisor will notify the chain of command and assure a command post is established distant to the site. A senior law enforcement supervisor or other manager should serve as Incident Manager and make other appropriate notifications. All security and law enforcement personnel on the site or venue will be alerted by a pre-determined code word transmitted by radio, telephone or word of mouth. All security and law enforcement personnel will immediately check their areas of responsibility for additional suspect items.

The fire department and EMS will be notified but not dispatched unless requested by the IM or bomb squad on bomb threats. The fire department and EMS should be dispatched if a suspect item is discovered and staged on site distant to the incident or at the nearest fire station.

The Command Post or security supervisor will clear a path and provide an escort for the bomb squad to the site. Usually a member of the bomb squad will serve as the Operations Commander (OC). The OC in concert with the IM will determine the best course of action. Typically a remote move, visual and/or fiber optic inspection, and/or x-ray inspection will be accomplished. If the package continues to be suspect or meets high-risk criteria, additional evacuation may be required up to 1,500 feet. Event cancellation and complete venue evacuation may be required.

A critical aspect is not to have tunnel vision during such an event. One should place officers looking outward for additional suspect items, persons whom are observing the event with interest, photographing or obtaining video of the event as the suspect item could be a test of the response to facilitate the bomber intelligence on how to defeat the countermeasures. The Command Post should not be placed in the most convenient location. Find the most convenient location and discard the location and pick another to defeat a pre-planned ambush or pre-placed IEDs. Make certain the Command Post is well secured and personnel are looking outward observing anyone in proximity to command post. Another consideration is for multiple devices, a hoax to distract law enforcement for the commission of another crime and parallel attacks at multiple locations. Bomb threats and some bombings are a distraction to consume law enforcement attention while a bank is robbed.

Actions Following a Credible Bomb Threat

- Confirm that the bomb squad is notified or dispatched per protocol.
- Consider the threat may be a distraction.
- Perform a search as dictated by protocol.
- Deploy as many security and law enforcement personnel as reasonable.
- Maintain secrecy of bomb threat and do not alarm the public.
- Security and law enforcement personnel should search with persons familiar with the area if time permits.
- Do not disturb suspicious objects or items that do not belong.
- Remember a device usually will not look like a bomb.
- Hasty searches should last no more than thirty minutes.
- Stay situationally aware and think multiple devices, distraction to commit another crime or a method to lure employees into the open for a firearms attack.

If a Suspicious Object is Found

- If an owner cannot be found within 15 minutes, evacuate the local area and call for the Bomb Squad. Following a bomb threat, evacuate the area immediately if a suspicious object is found and request the bomb squad, and at that point search for an owner.
- Move people away from windows in adjacent areas.
- If in the open, evacuate up to 300 feet initially (small objects such as a purse, small bag or box but evacuation of at least 1,500 feet are preferred) or a minimum of 1,500 feet for larger objects (brief-

cases, large boxes or like-sized objects.)

- Keep persons out of direct line of sight of the object.
- If you can see the object, it can hurt you!
- If object appears to be over 25 lbs. (briefcase or larger) evacuate to more than 1,500 feet and consider terminating the event.
- If a vehicle bomb, evacuate to 5,000 feet and terminate the event.
- Remember *time, distance* and *shielding.*
- Do not move or disturb the object.
- Do not turn on or off power, lights, utilities or equipment.
- Do not use radio transmitters or cellular telephones within 300 feet of the suspect item when possible. In some circumstances this may not be feasible due to the need for an expedited evacuation.

Explosions

- If the device explodes or if an explosion occurs, do not approach except to save lives.
- Evacuate persons through emergency exits in the area in an orderly fashion away for the device and if outside evacuate upwind avoiding smoke and fumes.
- Follow the Emergency Plans.
- Have the injured walk out of the explosion area and if outside walk upwind to an area designated which is accessible to EMS.
- Send the walking injured to a central location away from the explosion site, preferably a pre-designated EMS triage site.
- Have the injured that cannot move raise their hands.
- Enter the area only to expediently move the injured whom cannot walk and rapidly leave. Leave those who appear dead. EMS should use "load and go" protocols to remove victims from the blast area. EMS should consider removal of clothing to remove contamination form dust, debris, partially consumed explosives, toxic materials and off gassing of toxic combustions products. Removal of clothing will rid the victim of a substantial amount of contamination. Remember clothing may contain valuable trace evidence so one should know where the clothing was dropped.
- Remember a bomb may be used to distribute toxic chemicals, radioactive materials or chemical agents.
- Fight fires following explosions remotely and defensively. If no fire or injuries exist, leave the area to at least 1,500 feet away.

- The bomb squad may wait up to 20-30 minutes to approach the area and up to 60-120 minutes to enter the explosion site.
- Remember *Time, Distance* and *Shielding.* Stay out of line of sight to the explosion area and stay out from under glass.
- Remember secondary devices, ambushes and parallel attacks.
- Approach only after bomb squad performs a perimeter and site search clearing the area. Do not enter the explosion site to avoid destroying valuable evidence.
- Remember secondary devices that target public safety.
- Do not remove any items or debris from the explosion site as they may contain valuable evidence. Clothing from victims also may contain valuable evidence from explosions.

SUSPICIOUS PACKAGE OR OBJECT

Any one of these indicators may be the only warning you may have of a bomb. All are important. If the complainant feels the circumstances or item is interest is odd or questionable, treat the item as if it were a bomb. When in doubt isolate the item and take appropriate actions. Movement, opening, or a timer may trigger these items. It does not have to be large to be deadly.

- In an unusual place (box, bag, package, briefcase, purse, container, tape wrapped object, bottle with wick, wires visible or similar factors.).*
- Does not belong to anyone (particularly if in odd location).*
- Looks out of place (does not fit surroundings).
- Is tied with twine, rope or string.
- Package is heavy for its size or unbalanced if the package has been handled. Do not manipulate the package!*
- Unusual smell such as (fingernail polish remover or other chemical smell) or unusual container (fabric bag, wrapping, metal or plastic container.*
- Metallic sound or items loose inside package if the package has been handled.
- A stain on the package or the package is "wet" in spots.*

---

* Indicates a critical factor.

- Package well wrapped such as multiple layers of wrapping paper.
- Package has a large amount of tape on it, particularly electrical tape.
- Tape-wrapped objects such as bottles, other tape wrapped items.
- Do not move the package to an x-ray machine without bomb technician consent to observe the contents as some facilities have x-ray machines for search of incoming packages and personal items.
- If an object is found during a search of incoming packages or personal items the screener should stop the conveyer and question the owner. The item should not be disturbed if the screener reasonably believes the item is suspect. Local protocols should be followed but as a minimum the owner should be detained and searched, bomb squad called and the immediate area evacuated.
- Indicates a critical factor.
- Bottles or containers with liquid or a cloth or paper wick.
- Paper wrapped bottles or containers.
- Shipper or sender did not ship or send the package.*
- Package or item "just appears" location and does not belong to anyone.*
- Unknown method of delivery of a suspicious item.
- Exposed wires, visible metal foil, or bulge in the package.*
- Looks like it has been tampered with, opened and resealed
- Just looks "suspicious" or "does not look right."

Protective Actions for a Suspect Item

1. Evacuate the local area 300 feet initially and to 1500 feet when feasible.
2. Consider use of sandbags or water filled plastic garbage cans or other shielding around but not on top of the device.
3. Keep people away from glass. Glass is deadly in an explosion.
4. Open windows and doors if in a structure to vent blast overpressure.
5. Do not place anything on top of the suspect item.
6. If you can see the device, it can hurt you.
7. Keep people out of direct sight of the device.

Chapter 7

EVACUATIONS

In many instances when you arrive, you will find an evacuation has already taken place. This usually means that groups of employees are milling about public entrances. This is the most likely place for a bomb to be placed and is by far the easiest location to place a device. You should bring this to management's attention and if feasible move the employees to a safer location. The persons who need to search their work areas have already left and must re-enter the building to accomplish this. You may find many employees are reluctant to return and if the threat appears credible, this is not a wise decision. A roster should be used to confirm that all employees have been accounted for and their location is known. Officers will find it not uncommon in these circumstances to be unable to locate the person in charge of the facility and the individual who took the telephoned bomb threat.

If an evacuation has not taken place, usually the decision to search or evacuate lies with the person in charge of the facility. Know your agency's policy on this issue. The on-scene officer will be asked for his or her suggestion. If you elect to advise management to evacuate, have the employees look around their work areas for items that do not belong or are not recognized prior to leaving. They should report anything unusual to management. Have employees take personal items such as briefcases, purses, bags and related as they leave. Be certain that evacuation areas have been cleared of any suspicious items or person, as one scenario is to conceal a device outside the facility, make a bomb threat, and have a large number of employees gather and then detonate the device or ambush the employees with a firearm. Make certain that management has an accountability list to make certain everyone exits the structure. Employees should report to a central location established as a command post and any suspicious item they report should not be disturbed. If not instructed to not disturb any unknown item, employees may try to be helpful and bring the suspect item to the command post location. In some instances it may take longer to secure operating equipment such that the time limit provided in the threat may be exceeded. If no time limit is provided in the threat, you must assume that the device may explode at any moment. If evacuated, the structure should remain evacuated for at least 30 minutes to one hour past the specified time provided in the bomb threat. If no specific time is provided, these should be the minimum evacuation times.

A visual search should be made of evacuation assembly areas prior to sending employees to them. The bomb threat may be a ruse to get employees out of the structure. Once outside they can be assaulted with a firearm from a distance or a pre-placed bomb in the known evacuation area may be used. A management or supervisory employee should be stationed at each location to assure that employees are accounted for and do not return to the building prior to being released. Employees should scan their surroundings for suspicious items or persons, and report to a management representative. Cellular telephones, handheld or vehicular mounted two-way radios use should be limited within the structure and in proximity to it. However, the risk of low wattage radio or cellular telephone causing an improvised explosive device (IED) to explode must be weighed against the impairment of operations caused by this communications loss. Under no circumstances should any radio, cellular telephone or any other electronic device such as a metal detector be used in proximity to a suspect package. These devices may induce sufficient electrical current to detonate an electric blasting cap or cause electromagnetic interference in an electronic timer circuit causing a premature detonation.

If an evacuation is undertaken, the employees should remain away from the structure at the maximum feasible distance. Usually the quoted minimum distance is 300 feet with 1,500 feet recommended (see Figure 7-1). This may not be adequate if a particularly large device such as a vehicle bomb is suspected. Evacuation areas in proximity to parking lots are not recommended particularly if the parking lots are uncontrolled by security personnel. Evacuation assembly areas should not be located near glass. An explosion may shatter and propel razor sharp glass fragments at high speed or cause overhead glass to break

ATF	VEHICLE DESCRIPTION	MAXIMUM EXPLOSIVES CAPACITY	LETHAL AIR BLAST RANGE	MINIMUM EVACUATION DISTANCE	FALLING GLASS HAZARD
	COMPACT SEDAN	500 Pounds 227 Kilos *(In Trunk)*	100 Feet 30 Meters	**1,500 Feet 457 Meters**	1,250 Feet 381 Meters
	FULL SIZE SEDAN	1,000 Pounds 455 Kilos *(In Trunk)*	125 Feet 38 Meters	**1,750 Feet 534 Meters**	1,750 Feet 534 Meters
	PASSENGER VAN OR CARGO VAN	4,000 Pounds 1,818 Kilos	200 Feet 61 Meters	**2,750 Feet 838 Meters**	2,750 Feet 838 Meters
	SMALL BOX VAN *(14 FT BOX)*	10,000 Pounds 4,545 Kilos	300 Feet 91 Meters	**3,750 Feet 1,143 Meters**	3,750 Feet 1,143 Meters
	BOX VAN OR WATER/FUEL TRUCK	30,000 Pounds 13,636 Kilos	450 Feet 137 Meters	**6,500 Feet 1,982 Meters**	6,500 Feet 1,982 Meters
	SEMI- TRAILER	60,000 Pounds 27,273 Kilos	600 Feet 183 Meters	**7,000 Feet 2,134 Meters**	7,000 Feet 2,134 Meters

ATF I 5400 1 (01-99)

Figure 7-1. Evacuation distances for varying size vehicle bombs as recommended by the BATF.

and fall into the assembly area. Evacuation assembly areas should also avoid storage or operational areas that may house hazardous materials, flammable liquids, high voltage lines, or other equipment that presents a hazard if damaged or disrupted.

Many facilities disguise the evacuation as a fire drill. In fires, one factor is to control the spread of fire and smoke by closing doors and windows. In a bomb threat situation, the opposite strategy is needed. Doors and windows need to be opened to allow the blast overpressure to be vented and minimize structural damage. Security considerations in some facilities may make this impractical. One must also consider the security of proprietary materials contained within the facility such that some areas must be locked and secured.

In summary, a bomb threat should be reacted to based upon the totality of circumstances. A bomb threat made describing a device in detail occurring during a labor dispute coupled with a suspect package is credible. It should be fully reacted to as opposed to a simple threat saying ". . . there is bomb in a structure." The decision to search and evacuate should usually be made by the facility management with advice from the law enforcement or security officer. Know your agency's policy regarding your role in giving advice in bomb threat situations. Officers should not tell a facility manager that a facility has no bomb in it even following a detailed search by bomb technicians, electronic sensors and canines. The most detailed search may not find a cleverly concealed device. Reliance on your statement of safety may be the basis of litigation should the facility be reoccupied and a device detonates causing damage or injury.

Chapter 8

IMPROVISED EXPLOSIVE DEVICES (IEDs) AND EXPLOSIVES

Bombs are the weapon of choice for terrorists. Explosive materials are easily acquired or stolen, the devices are simple to assemble and deliver. Bombs are reliable and anonymous. They can even be delivered through the mail or a package service anonymously. Improvised explosive devices (IEDs) are sized upwards from a small pipe bomb up to a large vehicle loaded with explosives. Vehicle bombs have been effectively used to attack military, commercial and federal government targets within the United States. They are a favorite of European, Irish and Middle Eastern terrorist groups. Another tactic used with success in the Middle East is the roadside or command detonated IED. Once detonated additional devices are used to target those responding to the initial explosion or those attempting rescue the injured from the first explosion and such ambushes may be remote from the original event. Explosives may serve as dispersal mechanisms for other agents such as biological, chemicals, incendiaries or radioactive materials. Usually more than 2,000 bombings, or attempted bombings, occur in the United States each year with approximately 70 deaths as reported by the Federal Bureau of Investigations Bomb Data center. Bombings have decreased in number but the lethality of the attacks has dramatically increased. What had been limited to Europe, United Kingdom and the Middle East has arrived in the United States in the form of large vehicle bombs. The tactics used have also changed. Secondary explosive devices are a real threat as are chemical, biological, and radiological secondary devices.

Explosives are materials that form a transonic shock wave when they decompose. They are arbitrarily distributed into two groups with those that have deflagration waves at less than 3,300 feet per second classified as low explosive and those with transonic shock waves classified as high explosives. Filler refers to the explosive loaded into a device.

The faster the speed of detonation, the more shattering (brisant) an explosive is. Low explosives such as gunpowder typically must be confined such that the gases that form shatter the container rather than the shock wave. Low explosives deflagrate (burn) rather than detonate. Low explosives damage a target with shrapnel and a gaseous wave rather than the transonic blast wave produced by high explosives.

Explosives are further defined as primary, secondary explosives and blasting agents. Primary explosives are sensitive to fire, impact and shock such as lead azide, lead styphnate, mercury fulminate, and nitroglycerin, the homemade explosives hexamethylene triperoxide diamine (HMTD) or triacetone triperoxide (TATP) preferred by terrorists. HMTD and TATP are somewhat simple to synthesize but are unstable in the extreme being easily detonated by shock, friction or heat. Some primary explosives are used in detonators better known as blasting caps. These devices may be initiated by a burning delay such as a fuse or electrically. There exists another type of blasting cap known as non-electric that uses a flame impulse from a plastic tube known as a shock tube. Electric blasting caps and non-electric shock tube-type blasting caps are commonly found in commercial blasting. Homemade bombs however, can use a burning delay with commercial, homemade or hobby-type fuse. Hobby cannon fuse, normally green in color, is the more common fuse. Primary explosives include mercury fulminate and lead azide. The primary explosive in a blasting cap detonates and triggers a base charge of a secondary explosive such as pentaerythrite tetranitrate (PETN) that in turn detonates the secondary explosive main charge such as dynamite.

Secondary explosives contain explosives that are blasting cap sensitive but are not as sensitive to flame or impact as primary explosives. Remember all explosives including blasting agents may explode when involved in a fire. Examples include trinitrotoluene (TNT), PETN, dynamite, and cyclotrimethylenetrinitramine (RDX). Dynamite may contain a stable mixture of the primary explosive nitroglycerin that has been desensitized or other similar explosives (see Figure 8-1). Many dynamites use ammonia based explosives to render them safer for underground mining using an explosive which produces less toxic gas. The nitroglycerine based dynamites are now rare. Blasting agents are a compound that is not blasting cap sensitive. A secondary explosive booster such as PETN is required to detonate blasting agents. A common blasting agent is ammonium nitrate fuel oil mixture commonly known as ANFO.

The Oklahoma City and World Trade Center bombings were nitrate based substances (4,800 pounds ammonium nitrate fertilizer and 1,200 pounds of urea nitrate) sensitized with other agents. Common sensitizing agents include nitro methane and similar organic solvents. These bombs were designed to do maximum damage; however, their explosive composition was not optimal. Had they been military or commercial blasting agents, the damage and casualties would have been more substantial. Both efforts were the product of a significant amount of intelligence and planning. Both devices did an extraordi-

Figure 8-1. Old dynamite that is deteriorating. This is a common secondary explosive found in rural settings. It should never be moved or disturbed except by bomb technicians. Photograph by John Skipper.

nary amount of damage for their placement and size as one might note that the ammonia based explosives produce a substantial amount of gas which does the majority of the work of the explosive. These gases literally push and tear buildings apart during the explosion.

Blast overpressure is a two-phase phenomenon in which the short but intense usually transonic blast wave washes over the surrounding environment causing a jump in positive pressure. A second phase of relative negative or vacuum follows and is about three times as long but far less intense. The blast waves may be reflected from structures intensifying their pressures by pulses "stacking" on top of one another. The harder and less resilient the structure the more of the blast wave is reflected. This accounts for many of the odd damage patterns seen with explosions. The author has observed blast waves reflected from a nearby building overhang and causing injury to an individual some distance away from the blast.

Fragmentation comes in the form of primary and secondary shrapnel. Primary shrapnel comes from the container whereas secondary shrapnel comes from objects disrupted or damaged from the explosion such as glass. The primary cause of injury in large devices is glass. Glass may be propelled by the shock wave or broken and fall onto victims located a substantial distance away. Shrapnel may be dangerous to extreme distances. Speeds of shrapnel may be greater than 10,000 feet per second compared to high-velocity rifle bullets, which have velocities in the 3,000 feet per second range. Some bomb shrapnel may penetrate large concrete and nearby metal objects. Explosions produce blast overpressure which is a short-ranged phenomena and therefore less dangerous at a distance but shrapnel may be dangerous out to several thousand feet from the blast epicenter. To appreciate the threat posed by shrapnel consider the threat posed shrapnel as random high speed bullets shot in every direction from the bomb when detonated. One needs to be at least 1,500 feet away under and behind a substantial object capable of stopping a large caliber high speed rifle bullet.

Thermal events are the radiant heat and direct flames created by an explosion. The type material and configuration affects the amount of heat liberated and its form. This event is a bright flash or fireball which transmits a substantial amount of infrared energy sufficient in many cases to start fires at a distance. The fireball itself may also generate fires through direct flame contact. The thermal event may be enhanced with incendiary elements such as fuels or metal dusts. Fires are common following explosions. The explosives used in mining are typically over oxidized to minimize flame production but many military explosives are under oxidized to produce residual fuels in the explosion with a large thermal event and subsequent fireball.

Ground shock comes from interaction of the shock wave from an explosion with the ground. To be significant usually requires a large explosion such as the one used in the Oklahoma City bombing. It may affect the structural integrity of any nearby buildings in a fashion similar to an earthquake.

Explosions are the ultra-rapid oxidation of a fuel. Most explosives contain a combination of carbon, hydrogen, nitrogen and oxygen (CHNO) but TATP is an explosive which contains no nitrogen. The by-products in perfect reactions are: carbon dioxide (CO_2), water (H_2O), and nitrogen (N_2), but this rarely occurs outside a laboratory setting. The incomplete process of combustion produces carbon monoxide (CO), a toxic gas and carbon soot. If oxygen (O_2) is present, some nitrogen oxides may be formed, such as nitrous oxide (N_2O). The quantity of oxygen present intrinsically in the explosives determines whether the explosive is balanced with the appropriate ratio of fuel to oxidizer, or is under or overoxidized. The reaction products are formed in the following order with the first product nitrogen which forms nitrogen gas (N_2). The second explosion product, oxygen and hydrogen, forms water (H_2O) and the third product, oxygen with carbon forms carbon monoxide (CO). Carbon monoxide is a toxic gas. Any oxygen left then forms carbon dioxide (CO_2) from carbon monoxide (CO). Other trace products such as oxygen form gaseous oxygen and nitrogen oxides. Underoxidized explosives such as trinitrotoluene (TNT) and cyclotrimethylenetrinitramine (RDX) produce a large thermal event. They also produce large amounts of carbon soot forming black smoke. Overoxidized explosives such as nitroglycerine and ammonium nitrate tend to have white, gray or rust colored smoke and a smaller thermal event. These fumes indicate insufficient oxygen. Rust or reddish-brown fumes indicate the production of nitrogen oxides from the presence of higher than needed levels of oxygen. Nitrogen oxides are toxic gases. Some explosives such as lead styphnate or lead azide are characterized as pure explosives and contain neither a fuel nor oxidizer. TATP is an unusual explosive in that little heat is produced during its decomposition. The energy is released when chemical bonds are broken rather than by the traditional oxidation seen in conventional explosives.

HOMEMADE EXPLOSIVES AND EXPLOSIVE LABORATORIES

Explosives can be manufactured in a home setting or other non-commercial setting with relative ease. Those with the ability to cookbook manufacture methamphetamine are quite capable of manufacturing explosives. In fact differentiating an explosive laboratory from a methamphetamine laboratory can be difficult. Some differences might include the profile of the individual or individuals involved but explosives are found in a substantial number of the methamphetamine laboratories and the laboratory may have mixed application. Another concern is if the laboratory might be manufacturing biological toxins such as Ricin or chemical agents such as Sarin. These laboratories may have similar appearances to the laboratories manufacturing explosives and methamphetamine.

Home-manufactured explosives even have a similar appearance to methamphetamine in the crystalline form. One should be observant as most explosives laboratories may have a larger capacity and different chemicals present than one would expect in a methamphetamine laboratory.

One might expect the following items in an explosive laboratory:

- Electronics such as timers, circuit boards, wiring, batteries, switches, soldering iron, solder, duct tape and electrical tape, gloves
- Laboratory apparatus particularly more sophisticated glassware and heating apparatus
- Metal containers such as pipes, propane cylinders, pipe end caps or plastic container such as plastic pipe and end caps and PVC glue
- Chemicals such as strong hydrogen peroxide or pool chemicals such as strong peroxides or HTH (High Test Hypochlorite)
- Strong acids such as sulfuric acid or nitric acid
- Solvents such as acetone and MEK (Methyl Ethyl Ketone)
- Peroxides such as MEKP (Methyl Ethyl Ketone Peroxide)
- Hexamine such as contained in fuel tablets
- Benzene

- Urea products and urea-based fertilizer
- Ammonium nitrate fertilizer (34-0-0)
- Potassium nitrate fertilizer
- Sodium Chlorate
- Potassium Chlorate
- Sulfur
- Charcoal
- Glycerin
- Mercury compounds or liquid mercury
- Lead compounds
- Metal powders such as aluminum, zinc or magnesium
- Hobby fuse
- Fireworks
- Gun powder

If what is believed to be an explosive laboratory is found one should carefully exit the structure and isolate the structure, evacuate the immediate area and call the bomb squad. Never move what is believed to be explosives as homemade explosives such as TATP and HMTD are extremely sensitive to movement, friction or heat and may detonate without warning. TATP is in a constant state of decomposition emitting gases which may cause the container to fail and the resulting container failure may produce sufficient friction to detonate the remaining explosive. Never test what is believed to be an explosive or biological toxin with a drug test kit as the reagent may cause a reaction which emits toxic gases or worse, causes an explosion. Remember, explosive laboratories are very toxic, contain dangerous and reactive chemicals.

There exist two mechanisms of energy release when high explosives detonate. Rapid oxidation produces a shock wave and heated gases. Low explosives produce no shock energy as their effects are produced by gas energy. The higher the velocity rate, the greater amount of energy imparted, and the more substantial the damage or work performed.

In terms of energy liberated or work produced the shock energy accounts for a small amount of the work. Gas pressures perform most of the work. Shock waves dissipate rapidly over short distances whereas the gas wave can travel a long distance. Shock waves tend to shatter the material they interact with while gases push the material. This is why gas-producing explosives such as ammonium nitrate fuel oil mixtures (ANFO) blasting agents perform so well in mining applications where the object is to push material.

After the explosive event, explosion products such as gases, binders, and non-oxidized materials remain. Most of these materials are flammable and typically assist in the production of thermal event. Fireballs are particularly prevalent in under oxidized explosives such as TNT or RDX. In some under oxidized explosives, the thermal energy liberated is more than the energy released by actual explosion.

The carbon monoxide and nitrogen dioxide compounds produced by some explosives are very toxic. Entry into explosion areas should be delayed whenever possible to allow the gases and toxic combustion products to dissipate. Entry into confined spaces following an explosion should be made after atmospheric testing and following of confined space entry protocols. These protocols may not be feasible in emergencies where immediate rescue is required. The fumes and smoke from explosions provide several clues. Light gray fumes and smoke usually indicate complete combustion. Dark gray or black soot indicates an incomplete combustion with the production of carbon monoxide and soot. Many combustion products are acidic. Inhalation may produce lung injury. When exiting an explosion-contaminated area, a thorough decontamination effort should take place. No clothing worn in the contaminated area should be worn without appropriate decontamination. The personnel leaving the area should immediately wash their hands, take a shower and wash their hair thoroughly. Personnel working in proximity to an explosion site should remain upwind, uphill and not chew gum, smoke, drink or eat. One also must remember that explosives themselves are toxic and most impact the central nervous system, liver and kidneys. Persons who have been involved in explosions or contaminated with explosives should be decontaminated using copious amounts of water and soap when feasible.

Items should not be placed on improvised explosive devices (IEDs) or ordnance due to the possibility of initiation of the device by pressure. Placement of items over an IED or ordnance prevents the bomb technician from observing the item and may complicate the render safe procedure. If protection is essential consider expedient methods such as a sandbag containment of plastic garbage cans filled with water surrounding the IED. The ideal method is an upright bomb blanket which will absorb some of the shrapnel and blast energy. With bomb technician consent a foam-filled container may be used to mitigate the blast potential. The common techniques is to use a plastic bag designed for the purpose filled with an alcohol resistant aqueous film forming foam specifically formulated for this purpose. This will dramatically reduce the blast and shrapnel effects should the IED detonate. The author has conducted and published original research with one such device and found minimal blast effects from several pounds of dynamite at 10 feet. However, some shrapnel will still be projected albeit with minimal velocities. Several commercial devices of this type are manufactured but require special training in their application.

Another factor affecting energy transfer is the contact area of the explosive in actual physical contact with the object to be blasted. The larger the area of the explosive in contact with the material to be blasted, the greater the effect. Even small amounts of air between the explosive and surface to be blasted will cause a large degradation in performance. This is why standoff distance is so critical in protective measures.

The most dangerous type of IED used is the vehicle bomb. This allows a large amount of explosives to be transported by a vehicle to an optimal blast point. If structures have exposures such as the underground parking at the former World Trade Center or open front structure as in the former Murrah Building involved in the Oklahoma City bombing, vehicles can easily be sited for maximum destruction. Vehicles also have the ability to penetrate many barriers and overcome defenses such as the Marine Barracks bombing in Lebanon. Excluding vehicles from in and around structures most easily defeats these types of events. A standoff distance of at least 300 feet is recommended and longer distances when feasible.

Pipe bombs are constructed of metallic or plastic pipe. These IEDs are usually filled with gunpowder or other improvised explosive. They are reasonably simple to construct and deploy. They have a limited range and size due to transport and emplacement considerations. They produce significant shrapnel but are usually an anti-personnel device rather than any threat to a substantial structure. The common method of delay is a burning fuse, however, mechanical, electronic and booby traps are found. This is the most common type of bomb dealt with by public safety. These devices can be disguised or camouflaged in place.

Package and briefcase bombs are devices that are a closed bomb built into a container. The size limitation for hand carried bombs in this configuration is about 50 pounds total weight. This is an insidious device as it can be simply left and will detonate by a timer or anti-disturbance device. Mail or package carriers also may deliver smaller package bombs although the U.S. Postal service has weight limitations on packages not checked through a postal clerk. The container or package can be constructed to add to the shrapnel such as in the Other Side Club bombing in Atlanta, Birmingham Abortion clinic bombing and the Sandy Springs abortion clinic bombing which used metallic containers housing high explosives and nails.

Other types of IEDs can vary including booby trap devices, improvised hand grenades, rocket-propelled grenade, or similar devices. Items such as rockets, platter charges and mortars have been used in European and Irish terrorist attacks with success. However, technical expertise beyond what is normally required to construct a simple pipe bomb is necessary for these devices. A disturbing trend is the increase of command detonated IEDs that target first responders, suicide bombers, the use of secondary devices against first responders along with ambush with firearms, and the use of high explosives rather than low explosives as the filler in these devices.

Scenes are not secure following an explosion until they are thoroughly assessed by the bomb squad. A consideration for responders is the threat of a secondary device targeting responders, ambush using firearms, additional devices that have not detonated, suicide bombers and unconsumed explosives. Bomb threats are a common ploy to consume public safety resources while another location is attacked. An example is calling in a bomb threat to a school and then robbing a bank. In the ensuing confusion, a perpetrator may escape. If fire and EMS respond to a bomb threat, they should stage out of sight and no closer than 1,500 feet from the threatened structure. The policy of maintaining secrecy concerning a bomb threat is standard operating procedure for law enforcement and security to avoid "copycat" threats. Many jurisdictions do not transmit bomb threat calls for service except via land line telephones or encrypted radio.

Activities prior to the detonation of a device usually include searches, evacuation, and activities by the bomb squad and staging of resources outside the potential blast area. These are referred to as pre-blast operations. Following the explosion of an IED the activities are called post-blast activities. An example would be the assessment of the scene, treatment of the injured, rescue activities, firefighting if needed and evidence collection.

The size of the IED or suspect package will affect the size of the evacuation area and subsequent damage should the IED detonate. It will also affect other operations such as the number and type of injuries, location and size of staging area, request for mutual aid and threat level.

The exposures to blast damage will also be determined by the size of the IED or package and will affect the evacuation area size, location and size of staging area, request for mutual aid, and threat level. Structures that present high risk include those with high or dense occupancy levels such as shopping malls, stadiums, schools, or hospitals. Other structures with large amounts of glass, structures that cannot be easily evacuated such as hospitals or nursing homes, structures that contain hazardous materials or hazardous processes also present a high risk.

The protection of various structures will depend upon a variety of factors such as the type construction. Structures with hardened walls (masonry or reinforced concrete) are more resistant to blast forces than those made of non-masonry materials. Structures with large amounts of glass are very susceptible to damage and casualties from flying glass. One protective measure is the shatter resistant plastic coating on windows which minimize glass breakage. All responders and evacuees should remain out of sight and as far away as feasible from the threatened structure. Glass should be avoided particularly when overhead. Blast waves may follow unusual paths and may be reflected by terrain or urban features.

The number, type, and seriousness of injuries will depend upon the type, location, size, occupancy, and exposure of persons to the blast wave, shrapnel and glass. If the IED is large or detonated in a densely occupied area, the number of casualties can be large. Initial EMS responders may be overwhelmed by the mass casualty incident. The first on-scene EMS providers should ascertain the scope of the incident and request mutual aid prior to engaging in EMS activities. Triage is critical. Establish a triage area outside the blast area and if possible out of the direct line of sight. Remember secondary device and ambush with firearms and do not use the most obvious or convenient location. Have those injured walk to the triage area. Identify those who are still alive in the blast area. Use minimal resources to enter the blast area only to save lives. Minimize time spent in the blast area by "load and go" EMS tactics. Leave the dead. Fight fires with remote appliances and do not enter the blast area if possible. Use *time, distance,* and *shielding.*

Resources on hand are likely to be insufficient to deal with a moderate to large IED in a densely populated structure. Make certain that adequate EMS resources are dispatched and recalled. Fire fighting and EMS activities may have to be altered if unconsumed explosives, undetonated IEDs or secondary devices are found. Look to the bomb squad for advice in this regard. Under no circumstances should personnel other than the bomb squad, approach, move, touch, cover, or otherwise tamper with a device.

Operations involving explosions and suspicious devices may have to be altered as the response time of the local bomb squad might take an hour. This means that responders will have to take appropriate actions based upon their training and experience until trained bomb technicians arrive.

The commitment of responders to dangerous scenes involving IEDs or explosions should be well thought-out ahead of time. The risk-to-benefit equation should be carefully reviewed. If only property is endangered then one is foolish to risk the lives of responders to save an already heavily damaged structure. Fight fires remotely. Leave the dead in place. Remember secondary devices and ambush.

IEDs can be only one of multiple hazards at an explosion site. Structural damage may threaten responders. Hazardous materials may be released. Fires are common following explosions.

Those responding to bomb threats or explosions should especially be vigilant for unusual factors or situations such as:

- Prior threats or bomb threats to the location.
- An unusual item or container that looks abandoned or out of place.
- Abandoned vehicles that look out of place, or are in an unusual or sensitive location.
- Unusual odors or strong chemical odors.
- Unusual or odd devices, tanks, or attachments to utilities or near sensitive areas. This is of particular concern to large containers of hazardous materials.

- Booby traps, trip wires, packages with wires attached or protruding from the package.
- Obvious open bombs containing blasting caps, explosives, pipes or similar bomb-like devices.
- Suspicious items received in the mail or by package carrier.
- Individuals who appear to be suspicious or out of place, particularly if dressed inappropriately for the environmental conditions such as a coat in warm weather which might conceal an IED.

TYPES OF DEVICES

Bomb components include the items required to construct a device. A device is defined as the assembled components of a bomb, incendiary, chemical, biological or radiological bomb or dispensing apparatus. A device may also be items that are assembled to appear to be a bomb but are non-functional. A device normally consists of a container, an initiator, firing train and primary explosive or dispersion apparatus. The initiator starts the firing train that detonates the primary explosive. The primary explosive is usually contained in a blasting cap when detonated explodes the main charge. Gases, flame (thermal event) and usually shrapnel are produced. The container is the normal source of shrapnel. Containers must be used when gunpowder or other improvised low explosive mixtures such as sodium chlorate mixtures are used. Gunpowder burns rapidly rather than exploding producing gases. The build-up of pressure causes container failure and the explosion. Blasting caps may be detonated by flame or by electric means. Some simple bombs may have only a battery, wiring, blasting cap and explosive while other more sophisticated bombs may use microprocessor timers and silicon controlled rectifiers.

A booby trap is designed to detonate when activated by an unwitting person. It may use a trip wire, pressure release or pressure application, movement, an electronic detector, such as passive infrared or other electronic means to initiate an explosive device. These devices may be found in or around clandestine drug laboratories, marijuana fields, and are used frequently in domestic-related bombings.

Chemical devices usually produce some reaction chemically. The normal product is a toxic gas. A simple mixture of an acid with a cyanide compound will produce a lethal gas. Other chemical devices may use an explosive or other means to disperse a toxic chemical such as a nerve agent.

A closed bomb is a bomb or device built or covered in a manner to disguise it or prevent detection. This usually occurs when the device is concealed in a box, bag or briefcase.

Command detonation devices are detonated upon command of the bomber. The bomber may be off-site and use a telephone, pager, garage door opener or radio controlled servo to initiate the device. This type device although once rare has become common due to the proliferation of electronics suitable for the purpose.

Disguised devices may resemble something not associated with a bomb. This may include innocent appearing items such as a paper sack, flashlight, VCR tape, box or toy. The imagination and technical skills of the bomber are the only limitations. The covering container may be booby trapped in an attempt to prevent the contained bomb from being detected or rendered safe.

Hoax devices are items designed or built to look like a bomb. A hoax bomb is used to scare or terrorize and place fear in occupants of structures. These devices are excellent for the disruption of a business, school, or other event. Some real appearing hoax devices are sold commercially. Another common hoax device is a reproduction of military ordinance or de-militarized ordinance.

Incendiary devices are designed to start a fire. Some improvised explosive devices may be coupled with an incendiary to enhance the destructive effects. An example is a pipe bomb with an attached propane cylinder. Some are self-igniting if tampered with, such as a self-igniting Molotov cocktail. Some terrorist groups such as the Earth Liberation front use incendiaries preferentially. The ease of construction and minimal signature left by such devices make them appealing as fires have many sources of origin and the efforts to fight the fire may obscure the device components making evidence collection difficult.

A letter bomb is a closed bomb contained in a letter or envelope usually delivered by the postal service. This device is typically initiated by opening with a pressure release initiator or electrically with a bare wire loop.

Open bombs are devices that are not contained or disguised. These are easily recognized, as the components are visible. A related type of letter bomb is a package bomb, which the bomb is contained in a package, purse, briefcase or other container.

Pipe bombs are constructed around a pipe, usually metallic or plastic, but any closed container will work. Some pipe bombs will contain added shrapnel consisting of nails, screws, ball bearings or wire within the device or attached to its exterior. Fragments and shrapnel are the greatest threat from these devices since a low explosive is commonly used and overpressure is not a significant threat. Another hazard is the pipe bomb may contain spark-sensitive black, smokeless or homemade gunpowder. Pipe bombs are unpredictable and dangerous devices. This is most likely the type bomb a law enforcement or security officer will encounter.

Shrapnel may consist of augmented items such as nails, screws, and ball bearings or may come strictly from the bomb container such as a metallic pipe. Shrapnel may be housed internally in the device such as a pipe bomb or attached to the exterior of the device.

A vehicle bomb is installed within or on a vehicle. Traditionally, this has meant a device concealed in a vehicle targeting persons within the vehicle. This now may include the Middle Eastern or European model in which the vehicle is loaded with explosives and is designed to target persons outside the vehicle. This device can damage a large area while being reasonably inexpensive to manufacture. Some vehicles containing a bomb will be booby-trapped. The World Trade Center bombing and the Murrah building bombings used this type device. The term used to describe the device is a Vehicle Borne Improvised Explosive Device or VBIED.

INITIATION METHODS

This is the method whereby a device has its firing train activated and explodes. Various methods are explored. An uncommon method is barometric. This initiation method detonates with an air pressure change. Crude devices may use a balloon with expansion to bring wires into contact. More sophisticated methods may use an aneroid barometer. The usual target of this type device is an aircraft.

A collapsing circuit is a circuit that holds open a firing solenoid by battery or by electrical power alone. Any disruption of the power supply and this circuit closes. The closing initiates the firing train on an independent circuit. This is an unusual device and requires a high degree of sophistication on the builder's part. This circuit is designed to defeat interventions to render the device safe. The device will also explode once the battery power is exhausted on the collapsing circuit.

Command detonation is a device in which a person closes the firing circuit. This device can be used by a suicide bomber or a bomber who desires to be present when the device detonates. Frequently this bomber will use remote command detonation by a radio-controlled device such as a pager, cellular telephone, garage door opening or other simple radio frequency device.

Light causes the device to detonate. This device normally uses a photovoltaic cell wired to a blasting cap or solenoid circuit. These are rare devices used to detonate upon the opening of a package. However, several companies produce a product which alarms in the home when the residential curbside mailbox is opened to deliver mail and are adapted for used as an initiator with ease.

Penetration initiation allows detonation when a conducting substance such as a metallic knife blade penetrates package. The penetrating device completes an electrical circuit. Anti-penetration devices usually are two layers of foil. These are rare devices.

Movement-initiated devices are usually an anti-tamper mechanism such as a mercury switch, ball bearing in a cage or bare wire loops that close upon movement. These are common booby traps and can be used to prevent movement of the device or used to detonate the device upon movement. These devices normally contain a time to arm mechanism to allow the bomber to escape prior to the package arming.

The opening method of initiation releases pressure (mouse trap or clothespin contacts), or pulls bare wires closed to detonate device. This can be commonly used in letter and package bombs and is also found in booby traps. Trip wires use this method to initiate a bomb or firearm and are found in clandestine drug laboratories and guarding the approaches to marijuana fields.

Pressure applied can be found in booby traps that require stepping on contacts and closing them. It can also be used by applying pressure to a primer to initiate a charge (BB soldered to a shell primer contained in tube which detonates a center fire cartridge). A simple land mine-type booby trap can be made in this manner by placing a board over a tube holding a center fire rifle round with a nail under the primer. When stepped upon the board pushes the rifle round onto the nail which detonates the primer firing the bullet through the foot of the person stepping upon the board. These devices may be used to guard the approaches to a marijuana field.

Pressure release initiation allows pressure-holding contacts apart to come together and complete a circuit. An example is a trip wire hooked to a non-conductive material separating bare wires placed in the jaws of a clothespin. When the trip wire pulls the non-conductive material away, the wire contact completing the circuit detonates the bomb. An alternative is the trip wire may pull a fuse igniter detonating the device which ignites a powder train to detonate a blasting cap which in turn ignites the bomb.

Temperature initiated devices use a bimetallic strip and closes contacts upon certain temperatures being reached. Common thermostats would be useful in this regard. This is an unusual method of initiation.

Time initiated devices may use a mechanical clock, electronic timer, time fuse or chemical reactions. Mechanical clocks are common but electronic timers are also used extensively in this application although additional skills are required to integrate the electronic timer and probably use a silicon rectifier in the circuit to fire the device. Military time pencils use a chemical reaction to initiate a bomb.

Trip wires are used to initiate a device when pulled, cut or otherwise moved. This either releases a pressure release device (mouse trap, clothespin contacts), or pulls bare wires together. This is usually a pressure release device and is a common form of booby trap. Trip wires may be used to discharge a firearm.

Several methods of damage may be found using various devices. Explosive devices use disruption through the ejection of gases, thermal event and fragment travel. Chemical devices damage through direct action upon living beings through a toxic effect, denial of use or corrosive action on the environment. Incendiary devices deny use, damage through thermal effects and the production of toxic gases.

Irritant substances can cause the occupants to flee due irritant effects, smell, or toxic effects. These devices whether harmful or not can cause terror and an examples would be dispensing oleo resin capsicum or tear gas in an occupied structure. Even hoax devices can be effective in disrupting operations.

Pipe Bombs

Pipe bombs are the most common form of improvised explosive device (IED) encountered. They range from simple pipes with an end cap drilled to allow hobby cannon fuse to pass through to sophisticated devices with anti-tamper or anti-movement mercury switches with the pipe loaded with high explosives. They are assembled easily and the components can be bought over the counter in an anonymous fashion. Any type of closed container can be used for a pipe bomb, but typically galvanized metal pipe or less frequently PVC plastic pipes are used (see Figure 8-2).

Since low explosives such as smokeless gunpowder are more commonly encountered, the stronger the container, the more energy is released when it fails. For this reason and for its ability to produce shrapnel, metal pipes or other metal containers are preferred. Juveniles may use small diameter copper pipes filled with gunpowder since these are obtained easily and no significant mechanical alteration is needed. The ends of the pipe are crimped and the necessity of an end cap and drilling a hole for fuse is avoided. These devices, while crude, can produce a small explosion and shrapnel. Another common

Figure 8-2. A partially disassembled pipe bomb using a PVC housing and simulated high explosive in the form of dynamite. Photograph by Cpl. John Skipper.

Figure 8-3. A tape wrapped metallic CO_2 cartridge filled with smokeless gunpowder. This is an example of a small pipe bomb. Photograph by John Skipper.

form of pipe bomb is an expended metal carbon dioxide cartridge used to power pellet guns, paint ball guns or bicycle inflators. The cartridge is filled with gunpowder and the fuse inserted through the gas release opening. Juveniles may scavenge gunpowder from small arms cartridges or fireworks to serve as filler for these devices. While crude, these devices do produce a small explosion and shrapnel. Even if the device fails to explode, the burning powder will "rocket" the device some distance.

Pipe bombs may be "booby-trapped." They may use simple safety, hobby fuse or be on a mechanical or electronic timer. They range from open bombs that are easily detected or recognized to well-concealed pipe bombs in packages. Their size may range from small carbon dioxide cartridges used to power pellet pistols to a large diameter galvanized pipe several inches in diameter and more than a foot long (see Figure 8-3). Do not move pipe bombs as they may be initiated by pressure release, anti-tamper initiated or initiated by movement with a mercury-type switch or contain a friction sensitive explosive such as TATP. Many bombers use galvanized pipe as this produces shrapnel with some bombers adding nails or ball bearings to enhance the projected fragments. Most are filled with smokeless gunpowder, infrequently black powder, an improvised chemical mixture or rarely high explosives such as dynamite or plastic explosives such as RDX. Improvised mixtures may contain TATP or sodium chlorate.

Pipe bombs are never disassembled due to gun powder or explosive such as TATP potentially being in the threads of metal pipe and igniting upon turning or the pipe containing a pressure release detonator booby traps. On some occasions, a common hobby fuse may be used. Sometimes the fuse burns but malfunctions and the device fails to detonate. This is not a safe condition. Pipe bombs may be ignited through sparks, open flame, friction of movement of the end caps or sufficient static charge. This is particularly a hazard if black powder or an improvised chemical mixture such as TATP is used.

Pipe bombs may not look like pipe bombs. They may be packaged or housed in another container. Metallic pipe bombs may be located off center in a package making the package have an eccentric weight. This is a package bomb indicator. Avoid tunnel vision if a bomb is located. One device may conceal a second device. Always think secondary and multiple devices.

Anti-tamper or anti-movement pipe bombs may consist of galvanized pipe or PVC pipe, filled with gunpowder and shrapnel. A mercury switch, battery, blasting cap, nails or shrapnel can be contained internally or externally within the pipe bomb itself. This device is usually designed to detonate when moved. Pipes with end caps should be treated as a pipe bomb for this reason. Only a trained bomb technician following an x-ray can differentiate between a pipe bomb and a non-threatening capped pipe.

A booby trap initiated pipe bomb that may be found in packages or attached to a trip wire is a pipe bomb with a mousetrap attached. When triggered, the bale of the mousetrap snaps closed on a shotgun shell metal end cap with a BB or similar metal item glued to the shot shell primer. The primer detonates, initiating the smokeless gunpowder pipe bomb. PVC plastic pipes are commonly used for this type booby trap since PVC is easier to drill or cut than metal. Fragmentation is added by attaching nails or ball bearings to the surface of the pipe with glue or tape. This type pipe bomb can be inserted in a package such that the mousetrap closes during opening and the device detonates. This type pipe bomb is strictly an anti-personnel device aimed at the person opening the package.

Another type deployment is using a monofilament fishing line trip wire as an initiator. An alternative is to use a clothespin with wires attached to a battery and blasting cap or igniter with a non-metallic insulator attached to the trip wire between the jaws of the clothespin. The circuit to the blasting cap or igniter is completed when the trip wire pulls the insulator from the clothespin jaws. Igniters can be made from thin gauge metal wire, a broken flashbulb or flashlight bulb or model rocket igniters. Rarely, pyrotechnic igniters such as electric matches are used. These devices produce heat or flame to ignite the bomb filler making gunpowder a preferable substance since a blasting cap is not required to ignite a gunpowder-filled bomb.

Pipe bombs using trip wire initiation have been found in booby-trapped clandestine drug laboratories and near marijuana fields. These devices serve as a deterrent and notification if a perimeter is breached. For this reason extreme care should be used in working in and around these locations. Any intelligence indicating a particular location is booby trapped, the bomb squad should be utilized to clear the area. The data varies by region but some regions find as many as 10 percent of clandestine drug laboratories having explosives or booby traps present. Marijuana fields have a similar finding with a substantial number using booby traps to guard the approaches to the fields.

Pipe bombs can also be set up for time by using a mechanical clock or electronic timer connected to wires. A circuit is closed when a specific time is reached detonating the pipe bomb. Again, blasting caps or another type igniter such as a Christmas tree light with smokeless gunpowder can be used. Although still unusual an electronic timer with silicon controlled rectifier and battery can be used to initiate pipe bombs.

The more sophisticated the device, the more probable the motivation of the bomber. Sophisticated pipe bombs are usually found in terrorist-type bombings, domestic bombings, professional bombers using this device to kill, or a serial bomber. Crude or simple devices are more likely to be a one-time revenge bombing or juveniles experimenting. One consideration is that criminals may divert public safety resources to commit another crime such as robbery or burglary using a bombing.

Command detonation is a device that is detonated upon the command of the bomber. This is usually accomplished while the bomber has visual contact with the location to assure the victim will injured by the bomb. A telephone, pager or direct wire to the pipe bomb can accomplish this. With a fixed telephone, some older telephone ringers supply sufficient amperage to detonate a blasting cap. Otherwise, a battery with silicon controlled rectifier can be used to accomplish the task with pagers, garage door opener or cellular telephones. Radio-controlled servos used in model airplanes can be used to close a circuit and discharge a blasting cap or other igniter. Devices of this nature were once rare but are being used more frequently due to the miniaturization of the electronics, ease of purchase and reliability of the electronic components.

Bottle Bombs

Bottle bombs do not usually present as hazardous appearing devices. They usually are bottles, plastic or glass, filled with a caustic liquid and aluminum foil or other metal. Another common filler is isopropyl alcohol and High Test Hypochlorite (HTH) known more commonly as pool chlorine and rubbing alcohol. These devices will detonate after adding the components at unpredictable times. The device may explode seconds, minutes, or hours, depending upon the concentration of the corrosive material, the container and the amounts of reactants placed in the container. In one case, a bottle bomb was found and discarded, as the finder was not aware of what it was. The eight-ounce glass soft drink bottle exploded later and blew glass fragments more than 100 feet. Juveniles routinely experiment with these devices and use them for destructive purposes.

The most common items used as a corrosive liquid, which may include drain cleaner, hydrochloric acid, and lye with aluminum foil as the metal with another method of gas production the addition of HTH and rubbing alcohol. These chemical reaction generates heat and hot hydrogen gas. Hydrogen gas is an explosive. The usual mechanism of explosion is failure of the glass or plastic container. Some plastic containers will not fail until they reach the 100 pound per square inches of pressure. Reports of these devices reaching a high enough temperature to ignite the hydrogen gas with resultant fireball do exist.

These devices are very unpredictable and when they explode the hot corrosive liquid mixed with pieces of aluminum foil and container fragments. These fragments may injure property or persons within a diameter of 100 feet.

The information to produce these devices is easily found. Most high school chemistry classes teach a lesson on making hydrogen gas by reacting mossy zinc metal and hydrochloric acid. Other locations describing the construction of these devices include the Internet and clandestine bomb-making texts.

Law enforcement and security officers should research their applicable state and local laws to confirm that these devices are illegal, as they do not meet the traditional requirements for a bomb. These devices should not be disturbed if found but notify the local bomb squad. Be suspicious of any container that has a "muddy" looking liquid or aluminum foil within it. These devices may present as a bulged plastic soft drink container. If the device detonates, it will typically produce a gray-silver residue. Any vegetation will be brown from corrosive liquid contact. Container fragments may also be present. Beware of the residue and fragments are usually still covered in corrosive liquid. Care should be used in collecting and storing this evidence to prevent burns from the corrosive liquid and deterioration of the evidence container. The larger devices may produce small craters when detonated on soil or grass.

Using solid drain cleaners mixed with water produces the same effect if placed in a closed container. Another technique is to mix dry ice and water, which produces carbon dioxide under pressure and will cause container failure due to the high pressure.

When dealing with potential bottle bombs use caution, as they do not appear dangerous but can produce injury at close range. Remember they contain a corrosive liquid, which may be heated. Once mixed and in a sealed container, these devices may produce over 100 pounds per square inch of force which usually causes the container to fail.

COMMON BOOBY TRAPS

Intelligence is critical in booby trap situations. If the law enforcement or security officer is conducting a raid or search knowing that booby traps are likely present the plan for the operation must include the bomb squad. The presence of booby traps will affect the method of execution of the raid or search. Dynamic entries are usually not feasible if booby traps are suspected or known present. Bomb technicians will be needed to clear a path for search teams. Again, if booby traps are suspected, entry should not be made without bomb technician support and the bomb technicians should have input into the plans for entry and operational plans.

Most booby traps are simple. However, infrequently those persons engaged in serious criminal enterprises such as clandestine drug laboratories may have sophisticated detection apparatus and booby traps present. Common booby traps typically can use pressure release such as a mousetrap hooked to shotgun shell in a pipe bomb or a mechanical pulling action such as bare wires pulled together to initiate a blasting cap (see Figure 8-4). Others may use clothespin type contacts that complete the circuit when the insulator is pulled free with a trip wire. Anti-motion switches occur in a variety such as bare wires that are pulled together when motion occurs or mercury switches hooked to a blasting cap. A simple improvised motion switch is a rigid piece of wire erect within a circle of wire with motion bringing the two wires together completing the circuit. Another variation uses ball bearings loose in a container between two wires with motion closing the circuit. Pressure applied booby traps operate, for example, with a center fire rifle or shotgun shell designed to fire due to a ball bearing being attached to the primer and the device placed in a tube. When the tube is compressed by stepping on it, the cartridge fires the bullet into the foot of the person stepping on the device.

Other traps are initiated by turning power on or off. Light bulbs filled with gasoline or gunpowder, or triggering blasting caps in a refrigerator hooked to its light that operates when opened. Blasting caps and explosives may be connected to refrigerator lights, placed in VCR tapes, or use mercury switches imbedded within the explosive itself. Light sensitive photocells will detonate when exposed to light and complete a circuit as described previously. Drawers may be easily booby-trapped with bare wires or trembler switches. Mercury switches can be employed in any common device such as a flashlight, pipe bomb, package, bag, briefcase, or even concealed within a stick of dynamite. Common household objects may conceal a pressure release booby trap such as a bucket, book, box, aerosol can, rug, tele-

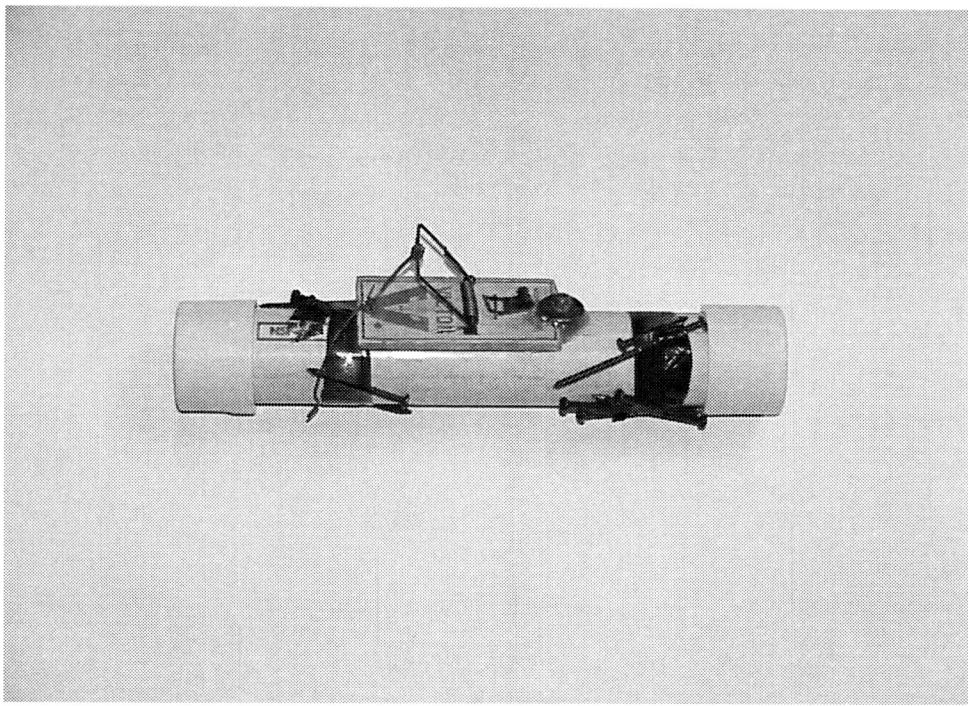

Figure 8-4. A booby trap using a mouse trap to fire a shotgun shell which in turn ignites the smokeless gunpowder contained in the PVC pipe. Note the added shrapnel in the form of nails. This device uses a trip wire to trigger the mousetrap. Photograph by Author.

phone, or other common item. Paths may have warning signs known only to the suspect such that unusual markings, strips or tape, surveyors tape, piles of pebbles or sticks may be a warning of a booby trap. When feasible paths to a clandestine drug laboratory or marijuana field should be avoided since the path is a logical location for booby traps. Remember booby traps may have the function of warning of approach, to slow the entry of law enforcement or discourage casual hikers or competitors from entry.

Molotov cocktails are easily booby-trapped. A mixture of gasoline and sulfuric acid is placed in the container with a textile wick. Within the wick is concealed a coin or other metal item such as a washer. When moved, the metal item drops into the acid-gasoline mixture and may ignite it in the worst-case scenario or generate enough heat to expel the mixture from the container. Another technique is to wrap the glass jar or bottle containing the gasoline sulfuric acid mixture in a

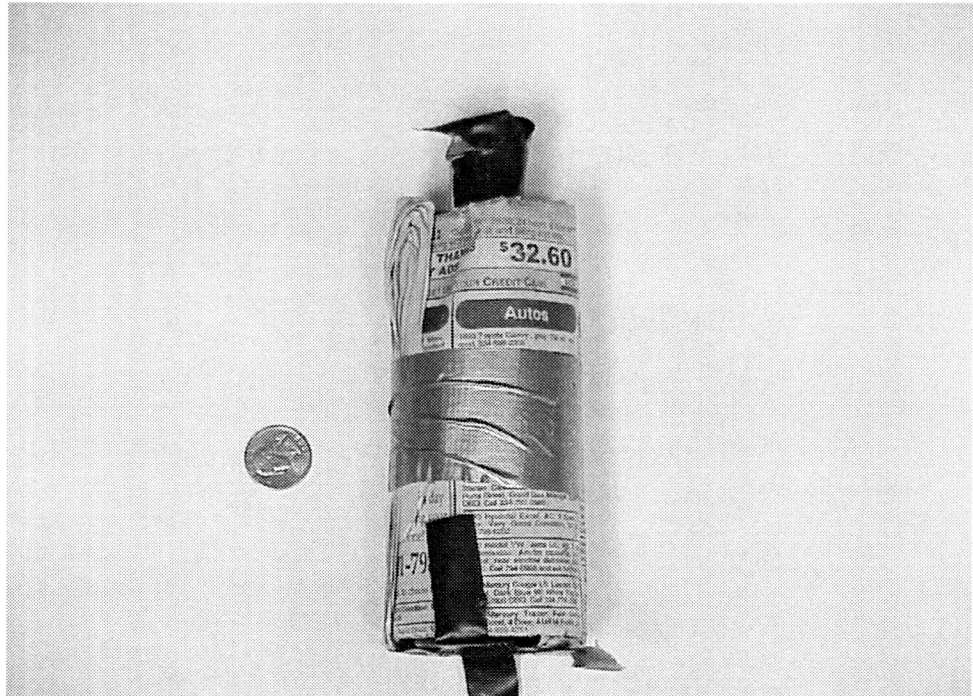

Figure 8-5. A self-igniting Molotov Cocktail incendiary device using newspaper saturated with potassium chlorate. A liquid mixture of gasoline and sulfuric acid is contained in the bottle. Photograph by Author.

newspaper saturated with sugar and a sodium chlorate mixture (see Figure 8-5). When the glass container shatters, the acid mixture should ignite the wrapper, which in turn ignites the gasoline. Care should be taken not to move these devices, as this hypergolic reaction will ignite if the chemicals are mixed.

Radio receivers can be adjusted in a manner to detect metal detector output and detonate. Metal detectors are not recommended for routine screening of packages. This same signal may cause spurious signals in an explosive circuit and detonate.

Hand grenades may have the delay train-type fuse removed and replaced with an instant smoke grenade fuse or a mercury switch. The grenade may conceal a pressure relief-type booby trap. Consider all smoke grenades and other chemical grenades as booby-trapped and do not move or handle them. One can simply acquire a demilitarized hand grenade, use a smoke grenade fuse, braze or weld the body closed and fill it with smokeless gunpowder. You now have a functional hand grenade. Even if the grenade has the "drilled" hole in the bottom, the original blasting, cap and fuse train may be in place. Consider all grenades as live until examined by bomb technicians.

Person-borne Improvised Explosive Devices (PBIED)

A person-borne improvised explosive device (PBIED) can be hand carried or concealed within the clothing. The maximum weight varies but normally 25 pounds is an optimal weight for concealment and explosive effects. The type explosives can vary but normally high explosives such as RDX or PETN are preferred but improvised explosives such as TATP are also useful. Low explosives such as smokeless gunpowder may be used but will have to be contained within a metal or plastic pipe in the form of a pipe bomb. Enhanced shrapnel will likely be used such as nails, ball bearings or other metal. Some efforts have been made to defeat metal detectors by using minimal wiring, plastic pipes and glass marbles to minimize the metal present. Fragmentation is the primary killer in suicide bombs.

Consider PBIEDs a smart bomb as a thinking person may be able to circumvent security measures. Responders or crowds of bystanders may be the target following an initial explosion. A second or third suicide bomber may attack the crowds, staging area, command post or even distant locations such as a hospital using confusion to defeat security measures.

One should also consider that the initial attack may be a ruse to cause evacuation and the main attack will occur along evacuation routes or upon responders to the initial attack. Always consider multiple attackers and multiple attacks.

No profile exists for suicide bombers but the following factors should be considered.

- Most are single young males aged 13 to 50 years but females have been used.
- Pregnancy clothing has been used to conceal a device and devices may be concealed in a bra.
- Clothing will vary but may be inappropriate, baggy, ill-fitting, long or bulky to conceal a device.
- Security, military, police, fire or EMS uniforms may be worn as a disguise.

- Look for a blank stare, nervousness or persons unresponsive to commands.
- Look for a stiff torso and the abnormal movements created by the device, or by switches sensitive to body position.
- Look for bulges, wires, switches or batteries, visible vests or belts under clothing.
- A backpack, suitcase, briefcase, radio, video camera or box maybe used to conceal a device.
- Look for wires or switches particularly if such is held in the hand.
- Shooters may be used to distract or neutralize security forces.
- If you see one suicide bomber think multiple suicide bombers.
- The initial attack may be a ruse to cause confusion for the main attack.
- The bomber may make a run to defeat security measures once close to the security barrier.

Security measures such as choke points with metal detectors and security searches and x-raying of hand-carried items are essential to defeat access by suicide bombers. However, remember such choke points are inviting targets in and of themselves. When feasible deploy a high ground observer and if a high-risk setting is present, consider deployment of a tactical precision shooter to cover such approaches. Use layered security such that if a suicide bomber penetrates one layer a secondary layer is available to stop the suicide bomber.

Clear rules of engagement for suspected suicide bombers must be developed and the personnel staffing security checkpoints and special events trained regarding the policy. The policy should be flexible such that persons meeting multiple criteria of a suicide bomber can be engaged with deadly force.

If a suicide bomber or VBIED is detected one should consider immediate evacuation and containment. The following issues should be considered.

- If a suicide bomber is suspected begin a containment of the suspected bomber.
- Consider immediate evacuation but remember the attack may be a ruse for collateral attacks of those evacuating.
- Consider immediate engagement with deadly force avoiding

torso shots using an immediately incapacitating head shot. Impacts on a device, worn typically in the torso region, may cause bullet sensitive explosives such as TATP to detonate.

- If negotiation is undertaken do so from a point of hard cover and at least 100 feet away. Make certain the suicide bomber is covered by another person with a firearm preferably a tactical precision shooter from high ground. Expect the suicide bomber to run to your position.
- Expect a suicide bomber to detonate the device if contained or accosted by security forces.
- *Do not close with a suicide bomber including those who are down.* Expect secondary devices, a remote triggering device, the suicide bomb may be booby trapped or detonated from a timer.
- *Remember, other suicide bombers may be among the injured.*
- If a suicide bombing has occurred increase security at hospitals as they will be a probably parallel target.
- Expect multiple suicide bombers.
- If a suicide bomber tries to surrender have the suspected suicide bomber show you his or her hands and keep them open. If wires or a switch is visible this likely is a suicide bomber.
- Do not concentrate all security forces at the initial bombing site or with an apparently surrendering suicide bomber as this may be distraction.
- If a suicide bomber attempts to surrender have he or she remove all clothing, shoes and turn 360 degrees. Do not allow a suspected suicide bomber to approach your position until cleared by the bomb squad if any devices are visible or any item is carried.
- Suicide bomb switches are likely held in the hands and are usually a push toggle-type switch or pull lanyard.
- Expect suicide bombs to be booby trapped such that removal, or mercury switches sensitive to body position can cause detonation, or a "dead man's switch" which causes detonation when released by the fingers may be used.

Suicide bombers are a problem for the foreseeable future. Only the use of protective intelligence, proactive procedures, layered security accompanied by flexible rules of engagement will defeat the efforts. Casualties during such events are probable and should be expected. Planning must address the early engagement of suspected suicide bombers with deadly force. The negative aspects of a false positive engagement with resultant death of a suspected suicide bomber should be planned as the likelihood of such an outcome is high as suicide bombings become more frequent.

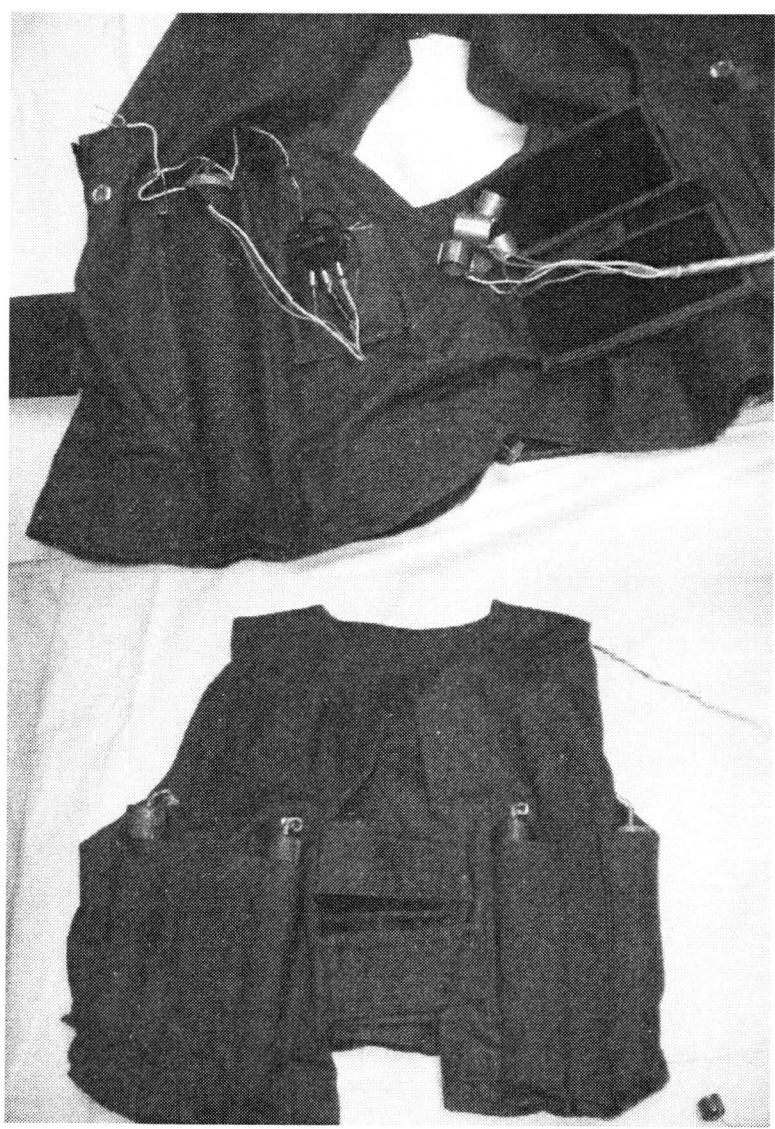

Figure 8-6. Suicide vest with wiring exposed including a toggle switch which can be pushed by the bomber to detonate, booby trap switch which will detonate the device if the vest is removed and finger switches worn as rings which will detonate if touched. The device contains sticks of dynamite with blasting cap initiators.

Chapter 9

VEHICLE BOMBS

The potential for vehicle bombs has not until recent years, been a strong consideration for law enforcement or security officers. With the current climate of terrorism, the utilization of vehicles to deliver improvised vehicle bombs or other weapons of mass destruction must be a consideration. Vehicle bombs were used to attack the Federal Building in Oklahoma City and the World Trade Center. Historically, vehicle bombs have been common in the Middle East and Europe but until recently an unusual event in the United States. According to the United States Department of Transportation Federal Motor Carrier Safety Administration, with more than 85,000 regulated trucks on U.S. highways, each potentially usable as a bomb. A typical passenger car can be loaded with more than 500 to 1,000 pounds of explosives. A large truck could contain upwards of 60,000 pounds of explosives. Those vehicles loaded with hazardous materials may not be dangerous from explosion hazards but may present threats from toxic or flammable materials.

The potential targets that might be of interest to those involved in terrorism include, but are not limited to, high occupancy sites such as hospitals, shopping malls, government buildings, stadiums, arenas or transportation facilities. Other sites of interest will be hazardous materials storage sites, infrastructure such as pipelines, bulk petroleum storage facilities, water storage and purification sites, bridges, communications facilities and similar facilities.

Law enforcement officers and security personnel should be especially vigilant concerning vehicles that appear out of place near high-risk locations. What in the past would result in a parking ticket now must be treated as a significant threat. Trucks that are placarded with hazardous materials and are out of truck routes or are in a location that is sensitive should be stopped, interviewed and investigated. Law enforcement officers certified as Department of Transportation inspectors will be invaluable in determining if violations exist. These officers are also familiar with the construction and operation of trucks along with the required documentation.

The best method for detecting a terrorist attempting to utilize a vehicle bomb or use hazardous materials to attack a location is increased vigilance and routine security and law enforcement methods. The in-transit time may be the most vulnerable time for detection but not for prevention of the attack. The vehicle bombs used in Oklahoma City and the World Trade Center used burning fuse delays. This produces a gray acrid smoke. Any vehicle parked in a sensitive location with visible smoke emitting from the vehicle or inside the vehicle may contain a bomb. Other indicators include, but are not limited to, stolen vehicles, stolen license plates, and lack of familiarity with vehicle by the operator, inadequate or improper written documentation, driver nervousness or unusual cargo.

Some facilities and locations may be barricaded to prevent the unauthorized penetration of vehicles into protected or vital areas. Policy considerations should be addressed as to whether the use of firearms or other deadly force is allowable should a vehicle attempt to evade a barrier or security officers to enter a protected or vital area. Attempting to disable a vehicle via gunfire is not reliable even with center fire rifle rounds. Efforts should be directed at preventing the vehicle from approaching the sensitive position with physical barriers resistant to crashes by large trucks. The use of bollards, concrete barriers, portable water filled barriers or similar devices are essential to protect valuable or sensitive locations. A minimum standoff distance should be 300 feet and more if feasible. However, many structures will not be able to achieve a standoff distance of 300 feet and may have considerations of an under building parking area or connected parking deck. In these circumstances vehicles must be carefully screened prior to access to such locations. Such screenings should as a minimum involve a physical search of the vehicle, electronic or canine detection of explosives. Remember substantial amounts of explosives can be concealed in non-visible locations within the vehicle such as between the vehicle exterior and frame, within the vehicle tires or fuel tank.

The next policy consideration is what to do if a suspicious vehicle is found in protected or vital locations. The level of suspicion will play a large role in the steps taken. Efforts should be taken immediately to discover if the vehicle is legitimate and has simply been parked in the wrong location. Inability to locate a driver or improper documentation should raise the index of suspicion. If the security officer observes smoke inside or coming from the vehicle, this may be the only indicator that a vehicle bomb exists. The only option may be to flee the location and to place a solid object between the vehicle and you while avoiding overhead glass or nearby glass windows or doors. Notification of the finding and evacuation of the area should begin.

Most incidents will fall in between these extremes where only a moderate to low level of suspicion exits. A well-written and concise policy should address the possible steps, which include:

- A vehicle that is in a protected area with a minimal index of suspicion.
- The inability to locate a driver in such incidents.
- If a driver is located, what steps should be taken to assess the documentation presented.
- Steps to take if driver cannot be located or if any suspicious indicators such as inadequate or improper documentation are present.
- When should an explosive detection canine or explosive detection apparatus be used?
- Evacuation versus removal of the vehicle or a combination of these tactics and what is the threshold for this action.
- Time constraints of what is acceptable.
- Criteria for bomb squad response.
- Discovery of what is believed to be a vehicle bomb and what steps should be taken including evacuation.

Interviewing a driver can provide additional information to base a decision. If the investigative stop is made in traffic, be cognizant of the position of the patrol vehicle and your proximity to traffic. Pick a good location for the stop if possible. Observe the license plate carefully and observe the driver for overall appearance, nervousness or impairment. Note any unusual chemical odors or smells from the vehicle. Question the driver about their trip and cargo noting any inconsistencies in the story. Determine if any hazardous materials are being carried. If this is a commercial truck, request the assistance of a DOT certified inspector if available. Examine the shipping papers for origin, destination and cargo type. Inconsistencies require in-route or cargo-type additional investigation. Check the documentation and tags for registration and wants.

Some out of the ordinary indicators for commercial trucks include:

- Trucks with explosives, chlorine or anhydrous ammonia out of truck routes or in densely occupied areas.
- Trucks with hazardous materials placards at a densely occupied location such as a shopping mall, stadium, government facilities, or residential areas.
- Trucks with poison gas cylinders such as hydrogen sulfide, chlorine, or other toxic gases near non-industrial settings.
- Gasoline tankers usually are on short haul runs of less than 50 miles.
- Rental trucks with hazardous materials placards or near a sensitive location.
- Driver with a Commercial Motor Vehicle license with no hazardous materials endorsement carrying hazardous materials.
- A truck marked with explosive placards that is unattended.

Perhaps the most difficult situation will be when a vehicle is located but the minimal criteria for suspicions are present. What is troubling is if a driver cannot be located or if the driver when located appears nervous or the documentation is suspect. An explosive canine may be deployed. An alert by the canine should be treated as a grave situation with a vehicle bomb present until proven otherwise. The security officer may have to proceed on intuition and experience. The policy should allow the unilateral declaration of an emergency by the ranking line officer present. If sufficient suspicion exists or if there appears to a vehicle bomb present, the immediate evacuation of personnel away from the vehicle, windows and overhead glass is essential. Removal of the vehicle to a less heavily occupied location if available should be a consideration; however, any personnel near a vehicle bomb are at substantial risk. Movement of the vehicle may initiate the bomb. Each facility has unique circumstances that must be addressed to include the potential for over or under reaction to a threat posed by a vehicle bomb.

Prompt bomb squad response for advice, evaluation and recommendations is essential. The bomb squad may attempt to render a safe procedure if there appears to be a bomb and lives are at risk, or they may be able to recommend a safe evacuation distance based upon their evaluation of the potential bomb.

The key to mitigation of a vehicle bomb is prevention. Heavily occupied sensitive structures and locations should be physically guarded and vehicles not allowed to approach within a minimum safe stand-off distance of at least 300 feet. This distance may need to be increased with consideration of the construction of the location, sensitivity of the occupancy or the potential size of a vehicle bomb.

Chapter 10

DOMESTIC BOMBINGS

The following information was derived from a study published by the Bureau of Alcohol, Tobacco Firearms and Explosives (BATFE). The profile of domestic bombings is different from bombings motivated by other factors. Domestic bombings may be classified as revenge motivation. Domestic bombings usually take place at the victim's home or workplace. Many domestic bombings occur at the residence of the victim. Usually one finds no hesitance on the part of the bomber to injure or kill uninvolved persons to achieve their goal. In fact, those relatives or associates who have been supportive of the victim may also be targets with their participation in support of the victim viewed as hostile behavior by the bomber. According to BATFE data, there have been 91 domestic bombings from 1974-1997 with 53 fatalities and 62 injuries. These attacks have a higher fatality rate when compared with bombings in general. The bombings are personal attacks usually well thought out with an above average construction of a pipe bomb using high explosives or gunpowder. These factors coupled with the finding the device is initiated by a "booby trap" leads to the higher fatality rate. Many of the devices are shrapnel enhanced and a few are incendiary enhanced.

The particular scenario found in these bombings is listed in terms of circumstances below. One or more may be present. The more factors present the more probable a bombing attack may occur.

1. Female-initiated divorces
2. "Bitter" divorce circumstances
3. A serious child custody dispute exists
4. A serious financial crisis occurs as a result of divorce

5. "Love Triangle" and not estranged or separated (infrequent)
6. Difficult property settlement
7. The bomber is usually a white male (rarely a female)
8. A restraining or protection order is in place through prior domestic violence or threatened domestic violence
9. The bomber does not like the former spouse's current partner

These bombs are typically well made and a large number utilize high explosives when compared to other bombings. These bombs have lethal results. The usual device profile is listed below:

- Well-constructed bomb is common
- Above average design of the bomb is common
- The device is a lethal pipe bomb
- The pipe bomb is 10 to 16 inches long may be up to four inches in diameter
- Many are placed in boxes and hand delivered disguised to look like mail or delivery package
- Many are disguised in a box and some are rarely disguised as an appliance
- Victim initiated detonation is by movement or opening the package
- Most use electric detonators
- Nine volt or a six volt lantern battery is usually the power source
- Smokeless or black powder is used in approximately 67% of devices
- High explosives are used in approximately 33% of devices
- Many are shrapnel enhanced
- Infrequently some bombs are incendiary enhanced
- Rarely command detonated devices are used
- Infrequently the bomb is placed in a vehicle

The bombing attack profile is as listed below. One or more factors may be present. The most significant is the utter disregard for human life and the lethal nature of the devices.

- Killing the victim is the only criteria
- Utter disregard for human life by the bomber including collateral injuries or death is common

- Well-planned attacks are usual
- The bomber may not care that his identity or suspect status is known
- The prime objective is to kill the estranged spouse or former spouse
- Infrequently multiple devices or multiple attacks are used
- Multiple victims are common due to the lethal nature of the device
- Rarely the victim's relative may be attacked
- The bomber may use a series of single devices but this is rare
- The bomber may use multiple devices simultaneously but this is rare
- Usually the attack occurs at home but some occur at the workplace
- The devices will usually be disguised as mail or a delivery package
- The device may be disguised as appliance or radio but this is unusual
- These are well thought out and thoroughly planned attacks
- Most devices are booby trap bombs requiring victim initiation
- Many are hand delivered but a few are mailed

The bomber knows victim's habits, profile and schedule making attacks simpler to plan and execute.

If an employee or victim is involved in a bitter domestic circumstance and is the recipient of a bomb threat or suspicious package, remember the domestic bombing profile. These individuals are at high risk for violent assaults or potentially a victim of a domestic bombing.

MAIL AND PACKAGE BOMB PROTOCOL

Personnel in the mailroom or other trained personnel should routinely screen items that are delivered by mail or other package services. The indicators for suspect packages or envelopes are numerous. Any one or more may be present; however, legitimate reasons exist that an envelope or package may meet the criteria. The most important indicator is the item is not expected. Items not expected should have the sender contacted to confirm the item was sent and is legitimate. If the sender cannot verify such, the package should be treated as a bomb.

Suspect Item Criteria

Any one of these criteria may be the only indicator of a bomb. If any one criterion is present, then consider the item suspect and isolate it. If investigation is unable to determine the package is, in fact, legitimate, and not a threat, then notify the bomb squad and consider evacuation. The United States Postal Inspectors are a valuable resource in such circumstances involving items sent by U.S. Mail.

- Item not expected or not requested.
- The listed sender did not mail or ship the package.
- Found in an unusual location or looks out of place, or "just appears."
- Item not delivered by the usual method.
- Handwritten, poor penmanship, misspellings, looks odd.
- Too much postage particularly if stamps.
- Postmark does not match return address or no return address.
- Envelope larger than pencil thickness (6mm).
- Envelope stiff or weighs more than two ounces.
- Restrictive labeling such as "confidential," "opened by addressee only," or similar markings.
- From out of country.
- The package is unbalanced or the package "sloshes" when moved.
- Twine used to secure package.
- Not professionally wrapped or has multiple layers of wrapping.
- Excessive tape particularly if electrical tape, or if metal foil is visible.
- Wires protruding or metallic sounds made when the package is moved.
- Package stained; wet appearing or leaking liquid, or an unusual odor present.
- Package looks like it has been opened and resealed or tampered with in any manner.
- Is from a known hostile person or group.
- Letter or package is labeled "anthrax," "Ricin," plague, yersina

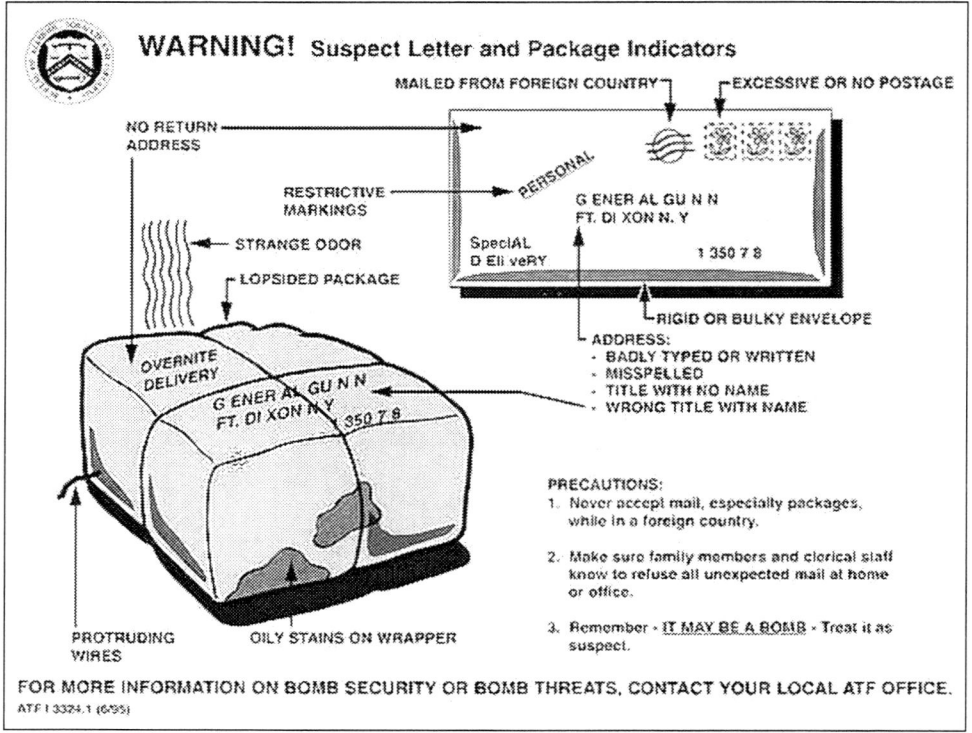

Figure 10-1. Mail bomb indicators as provided by BATFE.

pestis, or similar labeling indicating a biological threat.
• Letter or package has powder or dust-like material leaking from it.
• The package is marked with hazardous materials stickers.

Consider the mail and packages for those most visible in the organization as more suspect than routine items. Extra attention should be paid to these packages. The same holds true for packages addressed to persons involved in a controversy (see Figure 10-1).

Mail bombs are relatively easy to make and are anonymous to some extent. This device can allow the bomber to strike from a long distance with minimal risk to them. The U.S. Postal service now requires most packages to be sent in person. This may limit the size of the device and may make the bomber involve an unknowing third party but does not eliminate letter bombs. Letter or package bombs can provide a wealth of forensic data even if they explode. Crime scene protection is important (see Figure 10-2).

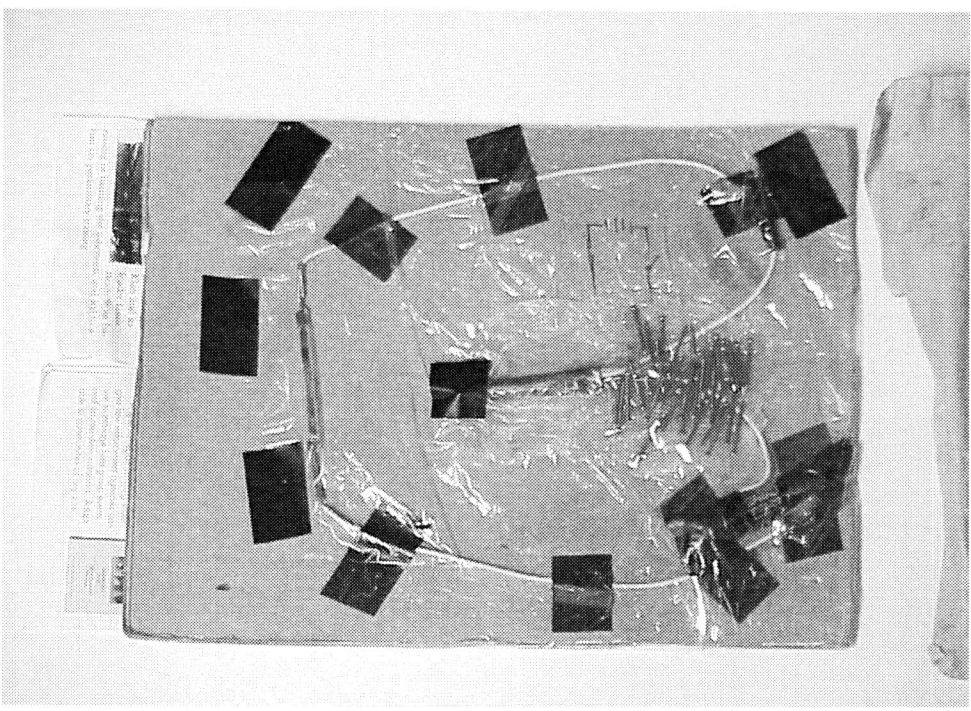

Figure 10-2. A letter bomb using a magnetic reed switch, simulated plastic explosive and metal shrapnel. Photograph by Author.

Threatening correspondence should be reported to local law enforcement. Threats of harm or the threatened use of force is a crime in most jurisdictions. Using the U.S. Mail to transmit these threats may violate federal law and the U.S. Postal Inspection Service can assist local law enforcement. The threat to use weapons of mass destruction is a federal crime and the Federal Bureau of Investigation can assist local agencies with such threats.

Chapter 11

INCENDIARY DEVICES

Incendiary devices are devices that utilize fire and flame to destroy property or injure persons. As a matter of federal law and in most jurisdictions, incendiary devices such as Molotov cocktails (flammable liquid in a container with an igniter such as a wick) are classified as bombs. This makes their possession for criminal use illegal and their use illegal. A substantial number of bombings in the United States involve the utilization of an incendiary device. The attack of a structure, particularly an occupied or potentially occupied structure including vehicles, with an incendiary device is a felony in most jurisdictions. The most common device used in fire bombings nationwide according to data from the Federal Bureau of Investigation is a container with a textile wick filled with a flammable liquid and hand delivered. This device is commonly called a Molotov cocktail. More exotic incendiary devices exist such as High Temperature Accelerants (HTA) which can be used to attack and destroy buildings with a fire that cannot be extinguished with conventional firefighting techniques.

The types of harm these devices may produce is a weak fuel air explosion and rarely a strong explosion. The most serious hazard is direct flame contact, radiant heat and structural failure due to fire.

These fires do produce toxic combustion products such as carbon monoxide, cyanide, acid gases and deplete oxygen. Most fire deaths are related to these effects and not burns.

Incendiary devices are easily assembled from household products, efficient in damaging property, leave minimal trace evidence and overall an effective terrorist tool. Fires can cause extensive damage and require significant resources to mitigate. The utilization of incendiary devices is increasing and a common event is to have incendiary devices coupled with conventional bombs. A propane cylinder attached to a pipe bomb exaggerates the thermal event and makes the pipe bomb more effective in causing damage. Many incendiary devices qualify under state and federal law as bombs. The Federal Bureau of Investigation's Bomb Data Center statistics indicate: that approximately 20-25% of all bombings involve an incendiary device.

These devices function approximately 75% of the time. Glass bottles with gasoline and a burning ignition source are the most common. This hand-delivered device is reasonably effective.

A variety of ways exist to classify incendiary devices. The usual method is by what actually triggers the device and how the device is delivered.

Triggers

- Burning Delay or Chemical Delay (most common).
- Electronic ignition (less common).
- Mechanical ignition (less common, found in booby-trapped devices).

Delivery Methods

- Hand delivered but not thrown.
- Hand thrown.
- Delivered by mail or other package carrier in disguised form.

Incendiary device components consist of an ignition source, flammable filler placed in some type of container. The ignition source is usually a burning delay such as a textile wick or may be hobby fuse. Some electrically initiated systems may use blasting caps or electrical components particularly model rocket ignition systems or electric matches. In some devices, the mixing of two chemicals (hypergolic reaction) will result in a delayed ignition. An organic liquid such as brake fluid or isopropyl alcohol and high-test hypochlorite (HTH) granules is mixed to form the hypergolic reaction. These items chemically react and depending upon the temperature and purity of the mixture may produce a delayed reaction and flame. The fillers usually consist of a flammable liquid. Gasoline is common; however, the substance may be solid material such as gunpowder or other flammable substances. The container is usually a breakable or frangible container such as glass are commonly used to disperse the agent once ignited or upon impact. This facilitates the spread of the incendiary agent and usually improves performance.

Common fillers may consist of roadway flares, military flares or flammable liquids such as gasoline, oil, and kerosene. Less common fillers include match heads, improvised gunpowder, fireworks and flash powder from fireworks, potassium chlorate-based compounds and other improvised chemical compounds such as HTA. The most common containers are glass containers such as bottles or light bulbs with an occasional plastic container. Propane and butane cylinders can be used when coupled with an explosive to allow explosive dispersal of the gases. This actually enhances the performance of the bomb and incendiary.

Several indicators of incendiary use exist. The same indicators for arson are utilized for incendiary incidents. Responders should be alert for fires in unusual locations or under unusual conditions. These incidents are crime scenes.

INCENDIARY USE INDICATORS

- Threats of incendiary use.
- Several sites within the same structure or multiple structures on fire.
- Accelerant burn patterns or the smell of accelerants.
- Broken containers or accelerant containers.
- Splattering of contents and resulting fire pattern from hand thrown devices.
- Residue from fuses.
- Indications of forced entry into the structure, particularly windows broken from the outside.
- Out of place appliances or other objects that are electrically powered.

A variety of methods can be used to detect an accelerant. Canines with the ability to detect accelerants are a common and proven technique. Responders may observe the broken containers or other indicators of incendiary use. Several chemical testing devices and electronic devices will detect accelerants as will trained canines. Remember, this is a crime scene and if an incendiary device similar to a Molotov cocktail is used, it qualifies as a bombing. The officer should remember a movement sensitive self-igniting Molotov cocktail is easy to assemble. Responders should not move Molotov cocktails or any sealed bottles found on a scene of what is believed to be an arson or bombing scene. Notify the bomb squad, as these devices may be booby-trapped.

AMMONIUM NITRATE

Ammonium nitrate fertilizers tagged "Oxidizer" can explode under the right circumstances. This is usually a "34-0-0" or "3400" tagged fertilizer. It has common agricultural uses. In the fire environment, these fertilizers are unpredictable. To assure the mixture is explosive, several techniques can be used. The mixing of a hydrocarbon such as kerosene, diesel fuel, nitromethane or any other organic or metallic powder sensitizes the ammonium nitrate mixture. It still requires a booster explosive such as dynamite or PETN to detonate it. Some mixtures of ammonium nitrate with nitroparafins such as nitromethane can produce a blasting cap-sensitive explosive mixture, which requires no explosive booster for detonation. Some binary explosives use the combination described above that is safe and non-explosive until mixed. Almost any organic or metallic substance mixed with high-grade ammonium nitrate will sensitize it, making it an explosive. The presence of ammonium nitrate fertilizer in the 34-0-0 grades may indicate a bomb making capability. Remember, an explosive booster is usually required to detonate this mixture.

Ammonium nitrate is formed from an ammonia and nitric acid mixture. When heated above 300° Fahrenheit the mixture undergoes physical and chemical changes that make the material unstable. The ammonium nitrate begins to decompose into nitrous oxide, water, and releases heat. This upward spiraling exothermic reaction can heat the mixture to its detonation temperature. With the right amounts of heat or mixing with other materials (particularly carbon containing materials such as sawdust, oil, diesel fuel, metals, or similar materials) will make it much more susceptible to detonation. The heat alone can cause it to detonate. Fire-fighting tactics should be from a distance with a deluge of water to keep the ammonium nitrate below 300°

Fahrenheit level. Remote appliances should be used and minimize exposure to personnel. Steam explosions may occur when melted ammonium nitrate encounters water or a fire stream.

Some fertilizer supply facilities and farm supply locations keep tons of ammonium nitrate fertilizer on hand. Remember, the higher the ammonium nitrate content, the more dangerous the material. Liquid forms usually are diluted and will run off, not presenting an explosion hazard unless heated in a tank. Then the hazard of explosion or catastrophic tank failure is possible. Contaminated ammonium nitrate fertilizer can be as powerful and destructive as dynamite. Ammonium nitrate is the primary explosive component of some military explosives and dynamites. Ammonium nitrate fuel oil mixtures are the most widely used blasting agents in many surface-mining operations. Smoke from an ammonium nitrate fire that is white indicates the ammonium nitrate has melted and is decomposing into nitrous oxide. Higher temperatures produce a reddish-brown smoke, which is indicative of high-level decomposition. All the fumes and smoke are dangerous.

The most dangerous form of ammonium nitrate is the "prilled" or wax coated pellets, which ignite around 300° degrees Fahrenheit. When ammonium nitrate approaches a temperature of 575° Fahrenheit, a second reaction begins in which nitric oxides are released in reddish-brown colored gas. The hotter the ammonium nitrate contaminated material, the more likely the explosion. If the reddish-brown smoke is visible, an explosion is probably imminent. The smoke is toxic. Some reports have been made of explosions in ammonium nitrate at temperatures of 800-900° degrees Fahrenheit. Train fires involving boxcars containing ammonium nitrate should be fought cautiously since the sealed cars make it difficult to gauge the extent of involvement of the cargo. Ammonium nitrate was the primary explosive in the Oklahoma City, World Trade Center (urea nitrate) and University of Wisconsin bombings that were expensive in terms of lives lost and property damage and all involved vehicle bombs. Ammonium nitrate fertilizer can be a powerful explosive due to the large amount of gases produced in its explosion. Ammonium nitrate is an excellent agent for cratering and building destruction and is used in mining for the "pushing of materials" though the large production of gases.

If a structure is well involved and it has ammonium nitrate, explosives or blasting agents inside, consider evacuation and probably not advisable to fight the fire. Protect exposures with remote appliances. If a vehicle transporting explosives or ammonium nitrate is involved and the cargo is on fire then evacuate at least 5,000 feet initially. Keep all personnel away. Remember, *time, distance,* and *shielding.* If it can be seen it can hurt you.

HIGH TEMPERATURE ACCELERANTS

High temperature accelerants (HTA) are related to solid rocket propellants. Several of the chemicals used in this mixture are actual components of commercial solid rocket propellants. These substances have been used in arson fires. HTA fires have been reported in Washington State, California, Florida, and Canada. HTA arson fires have caused several firefighter deaths. When used in sufficient quantity to start a fire the heat load rapidly develops, early flashover with rapid spread occurs leading to an early structural failure. This is true even in structures with little or no fire load. Water will not extinguish the fire and actually accelerates the burning. Errors in the composition of the HTA may result in an explosion of the accelerant rather than deflagration. Application of water to the burning HTA may result in steam explosions.

The problem facing bomb responders is identifying an HTA fire while in progress as the life threat to firefighters is high or discovering the incendiary devices prior to a HTA fire starting. The more serious scenario will be to identify an HTA fire while in progress. These fires have the potential to seriously injure or kill fire-fighting personnel.

These fires are reported to have bright white or orange flames. Pyrotechnic-like particles are frequently observed. Some officials have described the HTA burning as appearing similar to burning magnesium or an arc welder. The smoke is a grayish or dark gray color and pushes out of the structure forcefully. A rapid structural collapse usually occurs early into the fire. HTA fires are usually arson and may target an unoccupied building. Usually the building would be considered an arson target and suspect an HTA fire if an explosion is followed by an intense fire.

The dangers to responding firefighters include an extremely fast burning and spreading fire with early flashover. Extraordinary heat loads result even in buildings that have no significant fire load. In tests conducted by the Seattle Fire Department, one-hour rated walls failed within five minutes. If the fire is detected early firefighters may mount an aggressive interior attack and be overwhelmed by the extreme heat load. Another tactic used by the arsonist may be multiple HTA devices timed to progressively spread the fire throughout the structure.

Clues during a suspected arson fire that HTA might have been used are severe damage to the structure out of proportion to the fire load. Look for melted iron and steel objects including structural members. The Seattle Fire Department reported that temperatures in several HTA arson fires reached in excess of an estimated 2800 degrees Fahrenheit. Some metal items even may appear to have been partially vaporized. Expect severe concrete damage at the floor level in and near the suspected origin point or points of an HTA fire. The concrete will have lost its hardness and crumbles easily. Blue or green discoloration may be present. Some fires have produced crater-like depressions due to extensive damage by the HTA. Rapid progression of the fire and involvement of the entire structure with early structural failure are usually present. Witness or firefighter accounts of bright white flames, gray smoke pushing out of the structure, comments about pyrotechnic-like particles in the smoke or near the flames or arc welder-like flames are usually reported. The damage is out of proportion for the usual hydrocarbon accelerants. Laboratory tests are not usually conclusive for any accelerant, metal residue or oxides. These fires resemble combustible metal or thermite fires but neither metal slag nor metal residues are found.

In 1990, additional tests were conducted using an abandoned structure. The materials used to simulate the HTA were ammonium perchlorate, ammonium nitrate, eutectic salts, aluminum shavings and aluminum powder plus diesel fuel. The full-scale test using 462 pounds of the HTA in a 250-foot by 120-foot building managed to fully involve the structure with fire in approximately three minutes.

Like any mixture of fuel and oxidizer this compound presents a serious fire-fighting hazard as usual methods of removing the oxygen source or cooling below the ignition temperature will not work against an HTA fire. Essentially, once HTA is ignited only deluge quantities of water offer any real hope of containing the fire. The Seattle Washington Fire Department has produced a comprehensive report entitled "High Temperature Accelerant" that is well worth reading.

Chapter 12

PROTECTIVE MEASURES

Extensive materials are available in the public forum that clearly detail how to attack any structure, person or facility with improvised explosive devices (IEDs) or improvised incendiary devices (IIDs). Measures can be undertaken to protect and harden targets. Targets may be selected to make a statement to the public and usually the more visible the target the more likely to be attacked.

One of the initial steps in threat assessment is to read the material that is available to the public and particularly materials that might be used by threat groups. Knowing the common methods of attack will assist in developing a defense. Law enforcement agencies, particularly those with bomb squads, along with Emergency Management Agencies (EMA) may be of assistance. Be aware of any group or individuals that may have a grievance with the institution, facility, management, employees or clients. This includes: former employees, disgruntled clients, contractors, radical groups, activist or fringe groups, persons with legal or domestic complications within the institution, current employees with poor relations, employees or persons who may be mentally disturbed, and employees who have a domestic problem with a significant other who may target the employee's workplace. Employees should be encouraged to report problems that might affect the security or their safety in the workplace. Review past threats, and "hate mail." Reliable employees should escort probationary employees, visitors, and contract personnel when they are accessing any sensitive areas of a facility.

Work with law enforcement to review and gather intelligence to determine what groups may present a threat. Determine if the institution fits the profile of a target, these groups might be interested in attacking. Remember the more difficult a target is to attack, the more likely an easier target will be selected. Look at prior incidents; compare your institution with the targets and the results of the attacks. Think through the results and potential of success of a similar attack should it occur to your facility. Walk around the facility during all lighting conditions checking the exterior looking for any weakness or areas lacking adequate security. Look for points that are conducive to attack with a bomb or IID.

Hand-carried or person borne IED (PBIED) charges are typically limited to approximately 25 pounds. Exterior hand carried charges range upwards to 40 pounds (see Figures 12-1 and 12-2.) Vehicle carried charges can be much larger as a van can easily carry 2000 pounds of explosives. A 25 pound charge of high explosives under optimal conditions can do considerable damage not only to the floor the device is located, but to at least two floors above and below it. Fragment velocities can be in excess of 5,000 feet per second. A one-pound charge of high explosives will do considerable damage in a 30 foot by 30 foot room with masonry walls. A pipe bomb would produce fragment travel outside this radius if high explosives are used and a pipe bomb with low explosives would have most of the shrapnel contained by walls of typical masonry construction. Non-masonry interior walls provide minimal blast and almost no protection against fragments. Concrete or masonry floors in multi-story buildings may afford fragment protection if not disrupted by the shock wave or gas pressure of the explosion.

An exterior charge of 40 pounds placed in contact with a structure will produce considerable damage. However, any distance between the structure and explosive will significantly degrade its efficiency. This is why such is so critical that vehicles are kept as far as practical from structures as distance significantly degrades the ability of explosives to cause structural damage. Hazardous materials, valuable equipment and areas of high occupancy should not be located adjacent to exterior walls when feasible. High-risk areas should have at least two manned points of access (substantial locked doors) between their location and a public access point. An automatic fire detection and sprinkler system that covers the entire facility is a necessity. Monthly fire inspections by trained employees and annual inspection by the local fire department are recommended. If possible, have employees trained to form an on-site fire brigade with proper equipment for ini-

Figure 12-1. Approximately 30 pounds of C4 and TNT high explosives being loaded in a hole for destruction. This is the amount of high explosives that can reasonably be hand carried into a structure in a bag or briefcase.

tial fire fighting. This makes the facility more self-sufficient and reduces damages related to fires by prompt control and extinguishing. Make certain that personnel responsible for security consults the local fire department and have a fire preplan that is practiced annually. The fire department can provide valuable assistance in mitigating the fire threat through hazard reduction. Cross train the fire brigade and security personnel in basic emergency medical services training such that they can perform EMS functions until outside EMS personnel can arrive. The minimum level of training recommended is the Department of Transportation's 40-hour First Responder course.

If a detonation occurs several specific tasks should be undertaken to mitigate the incident. In the post blast environment involving a sizable bomb, the following can be expected: considerable structural damage; injuries from flying glass and fragmentation; injury from structural collapse with persons trapped by debris; people disoriented from injuries,

Figure 12-2. Crater created from the detonation of approximately 30 pounds of C4 and TNT high explosives.

blast and hearing damage; a dusty debris riddled environment with any open wounds seriously contaminated; superficial appearing wounds may actually be deep penetrations from missile impacts; loss of lighting and power; loss of emergency power to include failure of battery powered lamps; fire and flooding from sprinkler and water line failure; live electrical lines; release of hazardous materials; multiple casualties that overwhelm emergency medical service (EMS) resources; secondary bombs designed and deployed to attack responding public safety personnel; and loss of telephone, cellular and radio communications.

The post blast damage can be scaled to fit any device from small to very large; however, many of the above effects will be present even in small devices albeit in a more limited fashion. Plan for this event tailoring the plan around the best estimate of the most probable hazard the facility may face. Remember, a bomb does not even have to explode, or even be a real bomb (hoax device) to create terror and disrupt operations in a facility. Conduct "what if" drills with management, security, fire brigade and maintenance employees to include bomb threats, post-blast activities, medical emergencies, and other bomb-related incidents.

Exterior lighting and closed circuit television surveillance may enhance security of structures and facilities. The cameras should be taped on a professional grade video recorder, digital video recorder or optical media with the images retained for at least a 30-day period.

All windows and glass should have flexible plastic coatings applied to prevent missiles during explosions or high winds. Plastic impact resistant windows are recommended when possible. Glass is deadly in an explosion environment. Consider replacing glass with impact resistant plastic whenever possible. One point to remember is many plastics are sensitive to the ultraviolet spectrum in sunlight requiring the replacement of plastic exterior windows on a fixed basis. Those not replaced lose much of their resilience and become brittle with their ability to resist impact and penetration degraded. Follow the manufacturer's recommendation and use a professional glazier to install and maintain this type of security glazing.

Fencing with at least a 20-foot open gap between the fence and structure is recommended for minimally occupied structures. Barriers capable of stopping a vehicle with even more substantial distances are needed for structures that have a high-risk occupancy. Large reinforced concrete planters specifically designed for this purpose are esthetically pleasing when they contain plants or flowers, but are difficult to penetrate with a vehicle. However, they do provide a site for the emplacement or hiding of a bomb. Parking lots and vehicle parking should be located as far as possible from buildings. The minimum parking distance recommended for high-risk structures is 300 feet; however, this may not be easily achievable. Many structures have under structure parking and connected parking decks. Entry into these areas should be restricted and only allowed after physical search of the vehicles and detection by electronic or canine explosive detectors. Other portions of the structure that must be guarded, alarmed, surveilled or have restricted access include: access to high volume air conditioning systems, communications equipment, roof access points, basement or maintenance access points, utility connections, emergency generators and their fuel supply, ground floor entrances and windows, shrubs, and areas of concealment at the ground level. Access to the building should be monitored and allowed only after positive identification. Isolate the public areas and public access points from the essential or heavily occupied portion of the building.

Provide controlled photographic identification badges to all employees. Identification that utilizes a magnetic stripe, electronic "chip" or optical reader is significantly more difficult to defeat than simple photographic identification badges. The most complex identification is through the use of biometrics. This technique actually reads a physical feature of the employee such as their handprint, fingerprint or retinal image. These are difficult to defeat. Have a procedure for employees to immediately report loss or theft of their identification. Screen the public entering and inspect all items brought inside. Metal detectors should not be used on packages; however, x-ray screening is acceptable. Metal detector systems may induce sufficient current flow in electric blasting caps or electronic timers to prematurely detonate an IED. Screen all deliveries including mail.

Conduct bomb and security training for management, security, maintenance, clerical, and telephone operators. Have a simple, clear and concise written plan that requires drills with realistic scenarios annually involving everyone who might play a role in the incident. Critique and review all drills making changes as needed to the plan.

Conduct evacuation drills including bomb search of the assembly areas prior to evacuation. Designate assembly areas near the structure and assembly areas distant to the structure. Drill employees in bomb search techniques such that they can inspect their work areas prior to an evacuation. Should a device or suspicious object be discovered, have management recognize the necessity of using a distant assembly point. Provide procedures for a roll call and tracking of personnel should an evacuation be conducted.

Accountability is critical in a post blast environment. A minimum of 1,500 foot distance to the structure is recommended for an "away from the structure" assembly point. Make certain that management personnel can be located by responding public safety personnel. Train at-risk employees, particularly executives, as to vehicle and package safety. Suggest that vehicles be parked in a well-lighted and secure area with video surveillance. Consider a vehicle alarm system and remote vehicle starting system. Train the executive, his or her family and staff to screen their mail for the common bomb indicators. Conduct a security survey of their residences. A professionally installed home security system is recommended. Counter detection of surveillance is an important technique for executives to be familiar with. Alteration of routes and travel times should also be considered. Have a plan for call tracing when a telephoned bomb threat is received. Evaluate and decide what steps should be taken in response to bomb threats based upon the credibility of a bomb threat. If a limited search is to be implemented, have specific plans to include: exterior of the building, stairwells, and public access areas particularly restrooms. If an evacuation is required, evacuation routes and assembly areas should be searched prior to evacuation. The facility should have established plans and signals for the evacuation with employees inspecting their work areas prior to leaving in the event of a bomb threat. Avoid elevator use if possible. Establish a method for recall of key personnel with employees avoiding the use of radios and cellular telephones. However, if such devices are essential to operations utilize them sparingly. If a suspicious object is located, make a decision as to the actions to take but as a minimum evacuate the surrounding area. Do not move any suspicious object. The object or suspected device should not be congregated around or taken to an inhabited location. In 1917 in Milwaukee, Wisconsin, a suspicious package was taken to the Police Station from the Post Office where it exploded as it was examined killing nine police officers and two civilians. ISOLATE THE PACKAGE. KEEP THE CURIOUS AWAY.

The most important aspect of any bomb incident management program is that the procedures are simple and well thought-out. Management support with realistic drill scenarios is essential. Should an explosion, fire or other exigent situation occur, employees should assemble at pre-designated points. If a search has not been conducted of the assembly point, do so immediately. Conduct accountability roll calls and determine if any employees are missing. Send the ranking management figure to the public safety incident management post. Information regarding personnel should be routed to this individual. A person or persons who can communicate with the manager in the assembly areas should also be present. Make certain that public safety personnel are informed of the unaccounted employees and their potential locations within the structure. One issue to remember is a bomb threat may be a ruse for attack of employees once outside the structure or to gain access to the unoccupied structure once an evacuation has occurred.

EMERGENCY READINESS PLANS

Most emergencies are unpredictable limiting preparation. Occasionally other emergencies such as hurricanes may have some lead time for preparations. If preparation for the emergency is feasible, an increased readiness system is required to alert personnel for potential recall. One method is the use of "Readiness Conditions." These conditions require varying degrees of preparation by the agency. These are usually numbered in ascending order with Condition One normal operations. Condition Two indicates no potential for any emergency but the emergency is not imminent. Typically, the emergency is expected in more than 48 hours. Condition Three indicates an emergency is imminent and is expected in less than 48 hours. Condition Four means problems expected in less than 24 hours with final preparations rushed to completion. Condition Five indicates the emergency is in progress. Each law enforcement or security unit or specialty team should have the ability to increase their readiness due to specific threats. This allows preparation for responses that may affect that particularly unit. A summary of readiness conditions is listed below.

- Condition One–Normal Operations
- Condition Two–Potential problems or emergency conditions are expected in more than 48 hours. Readiness begins with notifications and testing of emergency equipment and inventory checks. The command structure is notified and resources audited. Test emergency generators, communications and assure emergency equipment/systems are functional. Acquire needed emergency supplies.
- Condition Three–Serious emergencies or circumstances expected in less than 48 hours. Rush preparations and assure that widespread notifications have occurred and the staff is ready to respond. As needed, place personnel on a short recall time.
- Condition Four–The emergency is imminent. Rush all preparations to completion. Begin the recall of personnel as the predicted emergency nears.
- Condition Five–Emergency in progress. Deploy the needed resources and consider mutual aid as needed.

Recall of specialty teams needs to be clear and concise in written plans. Field supervisors must be aware of the appropriate situations to deploy various specialty units. Their role should be addressed procedurally in writing to assure that the interface and chain of command is well defined.

Plans must address the following hazards such as severe weather and natural phenomena. Examples are hurricanes, tornadoes, flooding, drought, conflagration, blizzard, ice storms, earthquake, heat emergencies and volcanic eruption. Other emergencies such as hazardous materials releases, fires, sustained power failure or loss of other utilities, loss of communications, aircraft crash or cyber attack should be addressed in the plans. Other law enforcement and security emergencies include riot, mass shootings, mass casualty incidents, and hostage or barricade situations and weapons of mass destruction events.

Notifications during or prior to each emergency should be well planned. Alphanumeric pagers, text messages, automated telephone message units, emails or a manual telephone "tree" can accomplish the recall of personnel. A few agencies directly notify an off-duty supervisor who calls in his or her assigned personnel. An easier method is directly paging, texting or emailing personnel with their assignment. The personnel acknowledge the message as they report to duty. Obtaining adequate personnel is crucial early into the emergency as a lag time between notification and reporting to duty exists. This may be complicated due to weather or other conditions making travel to the facility or emergency scene difficult.

Unusual events may mandate special efforts as any location with a large or densely populated event has the potential for an attack. Emergency preparations should address the issues of crowd control, mass casualty incidents, crimes, riot, bomb threats, and fires. The orderly evacuation of the sites has to be well planned since evacuations can result in fatal trampling.

One issue that is an important consideration for a long-lived event is the consumption of overtime pay and expenditure of monies for resources. The budget of many agencies can be rapidly exhausted during these events. However, this may not be an issue in the initial stages of the incident but it will become one as the emergency progresses. Good record keeping is essential if any outside reimbursement by the state or federal government should become available. One important aspect is pre-existing mutual aid agreements are required for reimbursement of outside agencies responding to the emergency. Incidents involving hazardous materials or another technological emergency may have legal recourse to compensate the agency for expenditures.

Agreements for emergency resources and mutual aid should be established in writing. The agreements must define the condition under which they are operable and what each agency expects from the other. Lines of communications and control must be clearly delineated. As a minimum, a memorandum of understanding should be in place; however, many agencies formally use contracts for this purpose. How expenses will be paid is a thorny issue that must be addressed prior to the emergency occurring. Some agencies will absorb costs for short-lived emergencies when they deliver mutual aid but fiscal issues need to be discussed regarding longer lived or resource intense emergencies.

All plans must address resource acquisition on an emergency basis. The source and method for obtaining logistical needs such as, generators, barricades, fuel, vehicle repair, food, water, shelter, portable toilets and similar items. Pre-existing agreements need be in place with vendors to allow the emergency purchase or use of these items. Normal purchasing procedures will have to be circumvented but fiscal control and accountability must be maintained. Members of the financial and purchasing authority should review and approve these procedures.

Unique issues and problems out of the ordinary will happen. Written procedures should allow deviation to meet and mitigate unusual events. Exercises with probable incidents are necessary to test the plan and acclimate responders to potential emergencies. These events should not be so mundane that they do not require innovative thinking or solutions.

Chapter 13

OPERATIONAL SECURITY

Many law enforcement and security agencies take a nonchalant attitude regarding operational security. However, with the environment of terrorism and reasonably competent criminal elements, operational security is a valid concern. All facets of the organization should be reviewed.

Law enforcement and security operational methods and sources must remain confidential. Television with many reality programs that relate to law enforcement and security have been an "education" for the public but at the same time compromises critical security and law enforcement practices. All one needs to do to see the common practices of security and law enforcement in emergencies is to watch television. This includes responses to high-risk calls and the manner that intelligence is collected and utilized. Methods of conducting forensic investigations and the potential counter measures are also readily available.

What is the source of information that can be gathered about your agency? In many circumstances, press releases and media stories present a fertile source of sensitive information. Operations or accomplishments addressed by the media garner public interest but in doing so may compromise crucial data and methods of operations. What resources are available and how intelligence is gathered and who from the agency was involved in many instances is revealed to the media.

Extensive information about an agency, its resources, and special response teams is listed on their website. Review what is presented with caution in responding to requests for information about the agency. Some agencies proudly show their special response team. A careful observer can note how many personnel are available, what type of equipment is used, the type body armor in use and weapons available. Another potential problem is placing photographs of personnel other than command grade personnel on a website or in a public area. Undercover personnel can be identified from these photographs. Remember that non-encrypted email is not secure and subject to being intercepted. Sensitive material should not be allowed into non-secure areas such as a residence or on a home PC. Do not leave these materials visible in a vehicle or on your desk. Chat groups online can be another source of sensitive information since one never knows who is in the chat room.

Surveillance of communications is another common method of gathering data. The number of units in service, the call volume and other vital information can be garnered by listening to a radio scanner. Cellular telephones are not secure unless encrypted and sensitive information should not be discussed over them. A fallacy about encrypted radio communications is if the agency normally does not use them but suddenly begins to use them, this is a key factor alerting anyone listening that a special operation is in progress.

A simple giveaway of special operations is the gathering of vehicles with personnel donning tactical equipment. Out-of-the-ordinary meetings between the command staff are another good indicator as is the scheduling of additional personnel or support personnel. Try to avoid obvious schedule changes preceding special operations.

Most agencies discard important documents in the trash, which can be a source of information. Some agencies even allow inmates to handle their trash. All documents of security or law enforcement interest should be carefully shredded. Simple items such as travel requests, purchase orders or telephone message pads can provide valuable intelligence.

Security of documents is critical with plans and other sensitive documents marked as confidential. Vital information should not be stored on a computer outside the agencies control. Emergency plans contain rosters that can provide the names of personnel, their home address, telephone numbers and may identify members of special response teams. These simple documents can provide a wealth of information as to their composition and size. Data from the National Crime Information Center and state criminal justice information systems are very sensitive and must be carefully guarded.

Countersurveillance is a technique that law enforcement and security should take. It may allow the detection of surveillance by a group or individual who is intent upon identifying personnel or operational techniques. Some extremist groups have used this technique to identify agency personnel, their residences, personal vehicles, and family members. A poor technique is to hold sensitive meetings in a public place. Some agencies hold briefings and planning sessions in restaurants where sensitive information can be overheard by the public or employees. Officers should be alert into out-of-the-ordinary inquiries about their agency. Issues that are sensitive should not be discussed with family members or acquaintances.

The security of firearms, explosives and munitions by an unauthorized person is a great civil liability should the firearm be used improperly. These items should always be stored in a location that is theft resistant. Some agencies place vehicle alarms on their vehicles that have weapons stored while others use armored vaults in the vehicle trunk to store weapons. Bomb squads should have vehicle alarms and store explosives in theft resistant alarmed vaults.

Inmate labor is common in law enforcement and in many regards may be the only janitorial services available to an agency. Procedures must be in place to restrict their access to sensitive documents and areas. These are criminals who should be closely supervised while in law enforcement or security setting.

Many non-law enforcement agencies routinely operate with law enforcement and security. While many of their employees are trustworthy, it should be taken into account that in depth backgrounds and hiring requirements are usually not as stringent as those for law enforcement. One can find employees with criminal records working for these agencies. Extreme discretion should be used in sharing law enforcement sensitive material with other non-law enforcement agencies. Clerical and information technology personnel also have access to vital information. Background checks are needed to assure they are not security risks. Some fringe groups specifically target law enforcement by attempting to insert a person into a clerical position with access to sensitive data.

Some adversaries are observant and use indicators to assess agency capabilities. Drug traffickers and other sophisticated criminal elements may be intelligent enough to conduct counter surveillance. They can identify agency vehicles, personnel, by direct surveillance, websites and other open sources. The sources may include local news programs and the newspaper. Some have obtained building plans from open sources. Adversaries may use indicators such as the type vehicle, antennas, radios located inside vehicles to identify unmarked agency vehicles. The observations of arrivals and departures of personnel at the agency may indicate a special operation. The surveillance of radio traffic for an unusual volume or the use of encrypted radios can be an indicator of activities.

The Local Emergency Planning Committee (LEPC) and SARA Title III records are sources for the location, storage and protection methods of hazardous materials. The LEPC is important in coordinating response to emergencies including those that might involve the criminal release of hazardous materials. A broad spectrum of industry, public safety, media, and citizens may belong to this group.

Every agency must take countermeasures to include a comprehensive operational security plan that addresses how documents are handled and the methods to assure that sensitive information is stored. Secure communications must be provided along with regulations limiting sensitive information to only those "who need to know" for purposes of the operation. A comprehensive approach must be used to reduce the signature of security or police operations.

Chapter 14

BOMB INCIDENT MANAGEMENT
IN A HEALTH CARE FACILITY

Health care facilities handle large numbers of patients in stressful environments with potentially negative outcomes. Large numbers of employees also work in this stressful environment. This leads to issues that may have the facility; its employees, or patients as targets of threats or actual attacks.

Bomb threats and bombings have become a common method of attack on facilities and specific targets. Partnering with local public safety, security and the bomb squad can assist the facility to meet their needs for realism in scenarios. It can also meet the requirements of the Joint Commission on Accreditation of Health Care Facilities requirement for handling these emergencies.

These facilities are usually large and employ a large number of personnel. To add realism to the drills, each scenario should be unique and target a different area of the facility but require response facility wide of the bomb incident plan.

The hospital's plan should call for immediate notification to security of any bomb threat, threatening correspondence, in-person threats, or suspicious packages or items. The plan should include comprehensive training of all staff in identification and search for suspicious items. The security staff responds according to their plan worked out in concert with local law enforcement and the bomb squad. Notification of the police department is mandatory. When members of law enforcement arrive, a decision is made in conjunction with hospital management whether to implement the facility-wide Bomb Incident Management Plan or perform local searches or assessment in concert with hospital security.

If the facility-wide plan is implemented, all areas should be notified by a coded public address announcement, text messages, email and alphanumeric pager. The hospital disaster control facility should be activated and manned. Each operational division of the facility should be responsible for searching their area, followed by a report of findings to the disaster control center. Security and police officers should search the common areas and the exterior. Interior areas are searched by the facility staff occupying those areas. If suspicious objects are found, the area is secured and evacuated while the bomb squad performs assessment. Critiques are held following actual incidents and drills.

When a drill scenario is conducted, role players should be used and a variety of realistic scenarios can be performed. Telephoned bomb threats can be used to test the ability of the switchboard to make appropriate notifications and institute traces while testing the operator's ability to recall the specific threat information. A telephoned bomb threat can be coupled with a pre-placed training device that must be located either by security or other hospital employees. The security, law enforcement and bomb squad members in concert with facility management must then formulate a response.

The problem with a hospital environment is that the facility is heavily occupied and hazardous materials are common. Pressurized gases, toxic materials, radioactive materials and flammable liquids are common, coupled with patients that may be at risk if moved makes for a difficult at best environment for a suspect item. Some drill scenarios should force security, law enforcement, bomb squad and hospital officials to make decisions to attempt a hand entry render safe procedure with patients at risk, move the IED, or move critically ill patients from the danger area. IEDs should be placed adjacent elevators that could be used to move patients and from forcing the simulated evacuation down a stairwell. This will impose unique problems for hospital staff since many of the patients may be on mechanical ventilators or other forms of life support.

Other scenarios that can be utilized are threats received via mail or other methods of suspect package delivery to test the response in isolation of the suspect item and proper notifications. Chemical and biological threat/device scenarios can also been exercised to test the ability of the security, fire, bomb squad and hospital emergency department to deal with these simulated casualties. Domestic violence with threats of a bombing can also be utilized as a scenario to create a realistic drill setting.

One of the continuing themes should be realism and the unannounced drills should provide an environment that is not only stressful but test the ability of hospital management, security, and staff. These scenarios can test the ability of the bomb squad members. Law enforcement, fire and EMS should be familiar with the health care facilities in their response area as they present a unique response problem. All these facilities should have a bomb incident management plan to be accredited and bomb squads can provide a valuable resource in the writing and exercise of the plan.

Items that will be necessary to provide to security and law enforcement following a bomb threat include:

- Building and grounds plans
- A copy of the facility bomb threat response plan
- A briefing as to what steps have been taken prior to arrival
- Sequester any witnesses prior to law enforcement arrival
- Have witnesses write what they heard or saw
- Isolate any suspicious objects, packages and secure the area
- If evacuations have been instituted and their status
- An administrator or manager and security supervisor is present to coordinate with police
- A method to communicate with hospital security forces deployed is available
- Determine what additional resources are needed (fire, EMS, transportation of patients to another facility)
- Determine where best to stage additional resources as needed

Usually when the bomb squad arrives, they will assist in coordinating a search. Members of the bomb squad may or may not actually participate in searches. Some squads will only respond if a suspicious object or package is discovered whereas others will actively participate in the searches.

There exist several options when a suspicious object or package is discovered. If the suspect object or device is accessible and of low threat potential, bomb squad members may physically examine it. If the threat potential is moderate or high, an x-ray will be taken to disclose the contents. If it appears the device may be a significant threat, then several options exist.

If feasible, the device may be disrupted on site using high velocity water shot from a gunpowder operated cannon. Should the potential explosion of the device threaten patients who cannot be feasibly moved without substantial risk, then other techniques such as remote moving of the device to a safer location may be attempted. Another option would be to shield the device with a bomb blanket or other containment to attenuate shrapnel. As a last resort and only with lives at risk would bomb technicians attempt a hand entry render safe process.

Health care facilities should identify high-risk areas where patient evacuation would not be feasible should a device be located in those areas. These areas might include intensive patient care areas where vital life support equipment is used such as ventilators, operating rooms or similar locations. Other areas should have clear evacuation routes and alternative routes if normal routes are blocked or unusable. Horizontal evacuation is preferred, however, vertical evacuation may be the only alternative in some facilities. Remember, elevators may not be available or wise to use in vertical evacuations if a suspect device is located in proximity. Bomb-related evacuations could follow fire or other evacuation plans.

HEALTH CARE FACILITY EMERGENCY ACTION MANUAL POLICY & PROCEDURE

TITLE: Bomb Threat Procedure
EFFECTIVE DATE:

Table of Contents

Telephone Threats

Information Collection
Notification

Written Threats

Message Handling
Notification

Responsibilities

Administrator/Administrator-on-call
Communications Operator
Engineering
Security Services
All Departments

Search Procedure

Coordination
Marking Procedure
Actions if Bomb Found
Post-Search Procedure
Documentation

Bomb Information

Camouflage
Search Areas
Actions if Found

Checklist

Bomb Threat Checklist

Evacuation

Evacuation Procedure

TELEPHONE THREATS

In most cases, the Communications operator will receive the bomb threat call. However, any individual within the facility could receive the call. Whenever personnel within the facility have received a bomb threat call at any location, the following procedures to obtain information shall be used:

1. Listen attentively. Endeavor to ascertain whether the caller is male or female. Does the individual have an accent or speech impediment? Is he or she calm or excited? Is the call local or long distance? Ask the caller to repeat the message and write down every word spoken, if possible. Pay particular attention to peculiar background noises such as: motors running background music and other noises, which may provide a clue as to the locale of the call. Keep the caller on the line as long as possible and leave the line open after the caller finishes. Always write down the time that you received the call.
2. Immediately after the caller hangs up, the Operator will NOTIFY:
   a. Administrator/Administrator-on-call to relate information received.
   b. Associate Administrator/Nursing
   c. Director of Plant Services
   d. Director of Security Services
   e. Director of Safety/Pre-hospital Services

Note: If a hospital employee other than the operator receives the call, they should immediately call Security.

3. Since law enforcement personnel will want to talk first-hand to the person who received the call, that person should remain available until they arrive.
4. If the caller does not indicate the location of the bomb or the time of the possible detonation, ask for this information.
5. TREAT ALL BOMB THREATS SERIOUSLY; NEVER CONSIDER A THREAT TO BE A PRANK.

WRITTEN THREATS

1. Handle the message as little as possible after you have read it. Protect it so that it can be examined later.
2. Write down how you received the message, from whom, when, etc.
3. Notify the Director of Security Services immediately.
4. Notify the Communications Operator.

5. The operator shall then notify Administrator/Administrator-oncall, Patient Care Services Administration, Director of Plant Services, and Director of Safety/Pre-hospital Services.

ADMINISTRATOR/ADMINISTRATOR-ON-CALL RESPONSIBILITIES

1. Notify Police Department by calling 911.
2. Contact the President of the Medical Staff.
3. Request the operator to initiate "Code One" status.
4. If necessary, initiate evacuation of the facility or parts of the facility if it becomes apparent this alternative is necessary. Coordinate search effort.

SWITCHBOARD OPERATOR

1. Announce "Code One" ONLY as directed by the Administrator or Administrator on-call. IF DIRECTED, state "Code One" (pause) "Code One-Safety Drill, will all visitors and unauthorized persons please leave the building by way of the stairways." Repeat entire sequence one time at 30-second intervals.
2. Maintain an acute awareness of the "Code One" status in order to provide prompt attention to any support communication he or she may be required to render.

ENGINEERING

1. Engineering shall secure and lock all elevators.
2. Should report to the Administrator/Administrator-on-call to be assigned as a member of the search team.

SECURITY

1. The Director, Supervisor or designee should report to the Administrator to identify and access the situation.
2. Security representatives should be assigned to appropriate areas

to control entry by visitors and others attempting to enter the area of the bomb.

3. Security should admit or deny passage of visitors, depending on the nature of the visit and or business in the affected area.

4. Security should assist in identified areas as needed in search activities.

5. Security should assist in evacuation of identified areas when appropriate.

6. Security Representative, as well as other facility personnel, should utilize radio communication with extreme caution during bomb threats to reduce the chances of radio communication detonating a bomb. Radios should be turned off in identified areas.

ALL DEPARTMENTS

Department directors shall be responsible for the search of their respective areas under the direction of Administration. Assign two people to each search team. Each Unit Director will be responsible for the search of their unit.

SEARCH PROCEDURE

1. The Administrator/Administrator-on-call shall coordinate the search with local law enforcement authorities and hospital personnel. The local authorities are specialists in the field and should assume authority when they arrive but all efforts should be made through a cooperative effort with the Administrator. Since the authorities may not be familiar with the facility, an employee familiar with the area being searched shall be assigned by the Administrator/Administrator-on-call to assist in the search procedure.

2. When it has been decided that a search will be made, it should be made on a systematic basis. Each area checked should be marked with colored tape provided by security to indicate that the area has been searched and efforts will not be duplicated. First, you and your partner enter the room, proceed to, and stop in the middle, STOP, LOOK, and LISTEN. If the office

machines present are on, turn them off so the room is quiet. Both you and your partner should stand back to back as quietly as possible. If a clock is present or timing device is within the room, you may be able to hear it. Next, scan the area from the floor up. Look for anything foreign to the environment. A briefcase in the restroom would be an example of this. Next, divide the room in half and begin the search. The search is done in three separate sweeps. The first sweep is from the waist down. The second is from waist to eye level and the third is everything above eye level. When searching do not be afraid to be nosey. Check everything, but if you find something, its HANDS OFF. Report any suspicious items to the Security Personnel. When you report that the room is finished and no device was found, be sure that you feel good with that call. If you do not, search the room again.

3. If the caller identifies the area where the bomb is located, that area should be searched first. If no indication exists as to where the bomb may be located, a priority list will be needed if not all areas can be searched at one time. Priority should begin with the most critical areas and progress from there. If anyone finds what appears to be a bomb, THEY SHOULD NOT TOUCH IT. EVERYTHING SHOULD BE CONSIDERED A BOMB UNLESS DETERMINED OTHERWISE. They should clear the area and notify the Administrator who will notify the local police. The area should be isolated as soon as possible. Personnel should remain calm, so that patients will not become alarmed or excited and they should never indicate to the patients or visitors that a bomb threat has been received.

4. Open all doors and windows (where feasible) to minimize the primary damage from blast and secondary damage from fragmentation.

5. Remove all flammable or explosive material to a safe distance if possible, and if time permits. The engineering personnel will shut off the gas, water and oxygen in the area at the direction of the Administrator/Administrator-on-call.

6. After the area is searched thoroughly, the teams are to report to their respective department head that the area they were searching is clear. When the department head has received the information from all search teams that the department is clear of any strange or suspicious objects, he or she will notify the Disaster

Control Center, Administrator/Administrator-on-call, and will make his/her report after all areas are reported cleared to the Administrator/Administrator on-call. The Administrator/Administrator-on-call and local authorities should agree that all areas are "Clear," the PBX operator will be notified to announce "the Code One Safety Drill has been completed-please resume normal operations" (Repeat three times).

DOCUMENTATION

The searcher should report verbally to the Administrator/Administrator on-call after each step of the search. After the "Code One" status is concluded, each member of the search team should prepare a detailed written report to be provided to the Director of Plant Safety.

BOMBS

A bomb can be camouflaged in as many different ways as the imagination will allow. Some devices may be the size of a cigarette while others may be much larger. Some typical places to search are:

- Public areas (lobbies, waiting areas)
- Business Office
- Public telephone areas
- Lockers and filing cabinets
- In toilet tank reservoirs
- Behind and under sinks and plumbing, often suspended
- Basement areas
- Under stairwells
- Wastebasket and disposal cans
- Air-conditioning or heating vents and flues
- Around all types of equipment such as machinery, etc. These are examples only and by no means, the only places to look. Instruct personnel involved in the search that their mission is only to search for and report suspicious objects. DO NOT REMOVE JAR, OR TOUCH any suspicious object or anything attached thereto. The removal and disarming of a bomb must be left to

professionals experienced in explosives ordinance disposal. Bomb and explosives are made to explode, and no absolutely safe methods exist for handling them. Please allow our local authorities the opportunity to deal with these devices.

BOMB THREAT CHECKLIST

If you should receive a bomb threat, remain calm, listen to what the caller states, advise the designated hospital personnel, and complete the bomb threat checklist.

When you are talking to the caller, try to determine the following:

1. Where is the bomb located?_____
2. When will it go off?_____
3. What kind of bomb is it?_____
4. What is your name and address?_____

Try to determine from the caller the following:

SEX: MALE_____   FEMALE_____

AGE: ADULT_____   JUVENILE_____

ORIGIN OF INSIDE_____

                 OUTSIDE_____   CALL:

                 LOCAL_____   LONG DISTANCE_____

VOICE: HIGH_____   FAST_____

                 LOW_____   SLOW_____

                 PLEASANT_____   STUTTER_____

                 HOARSE_____   NASAL_____

                 RASPY_____   LISP_____

SOFT_____   OTHER _____

ACCENT: LOCAL_____   NOT LOCAL_____

FOREIGN_____   RACE_____

LANGUAGE: GOOD_____   FAIR_____

POOR_____   OTHER_____

MANNER: ABUSIVE_____   SPECIFIC_____

INCOHERENT___   RATIONAL_____

UNEMOTIONAL__   IRRATIONAL_____

BACKGROUND SOUNDS:

MUSIC_____   CARS_____

PLANES_____   STREET NOISES_____

TRAINS_____   PARTY_____

ANIMALS_____   LOUD NOISES_____

MACHINERY____   OFFICE_____

NONE_____   OTHER_____

OTHER INFORMATION: _____

_____

_____

_____

_____

EVACUATION PROTOCOLS

In the event that the Administrator/Administrator-on-call and local authorities agree that evacuation of a particular area is necessary, open all windows in area - close all fire doors manually-BE PREPARED TO MOVE PATIENTS by use of stairwells or certain elevators upon order of Administrator/Administrator-on-call. If some critical patients cannot be moved, place mattresses, or blankets around them for protection from fragmentation. The Police Department has expertise in the handling of explosives.

Central Safety Committee

Approved By:

Chapter 15

BOMB THREATS IN THE
SCHOOL ENVIRONMENT

B omb threats made affecting schools are a fact. All should be treated seriously. Officials should be particularly vigilant following bomb threats. Bombs and incendiaries are easily made from household items. School chemistry laboratories have the needed items to make bombs and incendiary devices. Small devices (tennis ball incendiary devices, 12 ounce drink bottle bomb) can maim and kill.

Bomb threats are likely to occur when high profile events are occurring such as tests, convocations or athletic events. A clear policy should be developed with school officials regarding security and searches. Prior to high profile events with large crowds such as an athletic event, a sweep and at least cursory search of the event venue should be conducted. If a bomb threat occurs secrecy should be maintained until a decision is made regarding evacuations. Preplans are a necessity as the orderly clearing a large crowd from a stadium, auditorium or gymnasium is difficult. In some cases with a non-credible bomb threat, the risk of evacuating may be more substantial than the risk of conducting covert searches during the event. Rehearsal of evacuation and search plans is important. The potential for panic and crowd problems should be a strong consideration particularly where a non-credible bomb threat is made.

If an evacuation is needed, evacuate only if evacuation areas are clear. The bomb threat may be a ruse to move students or staff outside for an assault or to detonate an explosive device preplaced in the evacuation areas. Officials should scan and preferably perform at least a cursory search of the evacuation areas prior to leaving the building.

If a search is instituted searchers should be alert for items that do not belong in a location, look out of place, or are unclaimed. Lockers are problematic and their search may require an explosive detection canine. Have students and faculty assure that everything in their classroom or work area is accounted for and no unclaimed items are present. If individuals are evacuated, have them take personal items such as book bags, backpacks, purses with them. Searchers should stay alert for items that do not belong at any location inside and outside the school.

Common items that should be of concern include sections of pipes of any type with end caps, any unusual item in an odd location, tennis balls in odd places. Capped soft drink containers with aluminum foil or "muddy liquids" in them should be considered as bottle bombs. Other items that may be suspect include unclaimed book bags, back packs or brief cases, grocery bags or boxes. If staff and students do not take personal items during an evacuation, there will be many suspect items, so many that it may not be feasible to clear them. DO NOT DISTURB ANY SUSPICIOUS ITEMS. Attempt to find out who brought or owns the item(s). If an owner cannot be located, consider calling for the bomb squad.

Most school bomb threats will be made via telephone. Have the school staff who are trained to handle these threats answer the telephone. This should include using *57 or call trace feature when a telephoned bomb threat is received. If these features are not available, do not hang up. Call 9-1-1 from another location, as a trace may still be possible. Written bomb threats or threats of any type should not be handled. Some threat notes or items may have fiber, DNA, and fingerprints obtained from the item.

Keep personnel away from glass. Evacuation areas should be no less than 300 feet from any suspect item. If you can see the item, it can hurt you. There should be a close and distant evacuation area. A distant evacuation area is recommended at least 1,500 feet away from the facility or preferably out of sight of the structure. An important requirement is to have a plan to transport the staff and students away from the facility.

Unusual packages with an unfamiliar sender or no return address should not be opened. They should be considered suspect and if a sender cannot be verified, the item should be isolated and the immediate area evacuated. The bomb squad should be called to assess the package. Any envelope or item that contains the words "Anthrax,"

"Ricin," powder, a threat, or other substance should be isolated. The immediate area should be evacuated and the bomb squad notified. Any threat that implies a chemical or biological agent is involved should be treated seriously. These usually are hoaxes but must be reacted to appropriately.

The faculty and staff should be alert to unusual student behavior. In many circumstances, students may be aware prior to a threat that an incident is going to occur. The faculty and staff should covertly observe student's behavior and demeanor. They may behave anxiously or erratically prior to incidents.

The facility staff should be encouraged to use rational judgment and do not panic. Most bomb threats are hoaxes.

SCHOOL BOMBER COMMONALITIES

The following commonalities were identified from a variety of open source materials. The sources included media and law enforcement publications. Much of the material is drawn from school shootings which had a substantial number also using explosives during the attack.

- More than 50% white males
- Ages 14–20 years
- Wore black clothing commonly during events
- Many associated with non-mainstream groups such as Goths
- Many considered loners or were members of fringe groups
- Most did not assimilate well socially
- Many were committed to suicide during or following the event
- Many threatened suicide prior to the event
- Most had a fascination with violence, bombs and firearms
- Most had access to firearms and bomb-making materials
- Most had impulsive and violent behavior displayed prior to the event
- The events were planned and not spontaneous acts
- Most made threats prior to the events
- Many had a fascination with violent figures such as Charles Manson or Adolph Hitler
- Many were fascinated by dark themes of death and suicide

- Many researched prior school shootings and bombings
- Some had mental health issues and some may have been depressed
- Most had been bullied or teased
- Many displayed anti-social behavior
- Normally some tragic event or personal failure triggered the attack
- The critical path was easy access to firearms and bomb-making materials

Some of the early warning signs may include:

- Social withdrawal or social isolation
- Stating he has feelings of persecution or "being picked on"
- Loss of interest in school activities
- Displays of poorly controlled anger
- Threats of violence coupled with rage and violent outbursts
- Suicidal ideation or suicidal threats
- Preoccupation with dark themes with morbid contents such as death
- Some triggering event such as a personal failure or problems at home.

Chapter 16

RESPONDING TO EXPLOSIONS

Law enforcement and security officers should treat all explosions as a criminal event until proven otherwise. One of the considerations is the location of the explosion. Care should be taken in responding to a location that is at high risk for attack. Always think worst-case scenario. Consider the structures and occupants and why they might be a target for not only hand placed, vehicle borne but also suicide bombers.

Look for out of place or stolen vehicles that might contain a secondary device, persons who might be a suicide bomber or those who might execute an ambush with firearms. Be alert to the surroundings and look for out of place items such as briefcases, boxes, bags or similar items that might contain a secondary device targeting public safety. Scan persons in the vicinity for appropriate behavior and clothing. Those wearing inappropriate clothing, trying to conceal themselves or who are watching responders rather than the event may be a threat. A small explosion may be a lure to bring public safety responders and citizens into the blast zone of a much larger device, to serve as targets for suicide bombers or for ambush with firearms. Explosions attract crowds of curious onlookers and the citizens who want to help.

Stay out of the blast area. Under ideal circumstances the bomb squad should examine the area working from the outer perimeter inward to test for radioactive material, chemical agents, clear it of secondary devices, unconsumed explosives, and booby traps while law enforcement establishes a secure perimeter to prevent an ambush or suicide bombers from attacking. This may not be feasible if the bomb squad is not immediately available and an injured persons whom cannot move under their own power out of the blast site. However, remember suicide bombers may be among the injured or dead with live IEDs.

Evidence may be projected beyond the area where damage is seen. A good indicator of damage from a blast is broken windows or debris in the street. If the blast area in which windows are broken is 500 feet, then the outer perimeter should be established at least 750 feet and the command post sited outside this area. A factor of 1/2 times the distance from the last found projected debris or broken windows to the center of the blast crater is recommended for establishing an inner perimeter of the scene. Remember to stay upwind and uphill when feasible. Select the best spot for a command post and do not use it. This is to avoid potentially pre-placed secondary devices targeted at responders, waiting suicide bombers or a preplaced ambush. Always examine any area to be used as a command post or staging area for secondary devices prior to use. Security of the command post and staging areas is essential. A hard perimeter should be established with personnel looking outward from the perimeter for threats. Resources may determine how big an inner and outer perimeter can be maintained. The perimeter should initially err to the side of too large such that the perimeter can be shrunk rather than expanded.

Keep the streets open for emergency vehicles but also positively identify emergency vehicles entering as being staffed by legitimate responders. The tactic of concealing VBIEDs, suicide bombers or armed attackers has been used with legitimate appearing or stolen emergency vehicles. Positively identify all persons entering the scene. Vehicular and pedestrian traffic will be a serious problem at most scenes. Establish staging areas away from the blast site and have perimeter security deployed. Have the injured whom can walk out do so and direct them to an EMS triage area. Law enforcement personnel should be present and consider a suicide bomber or armed persons may be concealed within the injured. EMS personnel should bear this in mind as they examine the injured and look for concealed explosives Limit the number of personnel entering a blast site to an absolute minimum. Keep the curious out including unneeded public safety personnel. Expedite the removal of the injured with "load and go" EMS tactics. Once the injured are removed, fall back to a location at least 1,500 feet or as far as operationally feasible from the blast site, preferably out sight of the blast location. Explosions may be a distraction for another event or used to disperse chemical or radiological agents.

Explosions are not effective to distribution biological agents as the thermal event from an explosion tends to incinerate the biological agents. If people are down from respiratory problems, having difficulty breathing or seizures, the responders should think chemical weapon, probably a nerve agent. Stay out of fumes, smoke, mists, liquids, or unknown substances. Site the staging areas and command post upwind and uphill of the explosion if feasible. Radiation and chemical agent surveys should take place early in the event.

The blast site should be entered only to save lives. Leave the dead in place. If fires are present, fight them defensively with remote appliances. The blast area should not be entered until cleared by bomb technicians if no injured are present to remove or fires to fight. If fires are small and contained, the risk of entering the blast area to fight the fires should be weighed against allowing the fires to burn.

Remember, the clothing of the dead and injured may contain valuable evidence and are probably contaminated. Some persons may leave an explosion site prior to the arrival of EMS and self refer themselves to medical facilities. The blast area will be rich in evidence that post blast investigators will need. Vehicles that were present in the blast area should not be allowed to leave until inspected for evidence. The number of casualties may overwhelm initial responders. Request mutual aid early into the event as it will take time for these resources to arrive. Structural collapse with entrapment is always a possibility. Urban collapse search and rescue resources are usually limited to larger urban areas and may take an extended time to arrive. Local resources may initially be "on their own" for several hours until outside resources from surrounding locations, the state or federal level can arrive.

If law enforcement personnel request assistance in collecting or handling evidence, remember that valuable forensic evidence can be obtained even following explosions. Items may contain explosive residue, fiber, fingerprints, and DNA. All items should be handled while wearing gloves to prevent cross contamination and to protect the wearer from body substances. Wet items should be placed in paper bags initially and then allowed to air dry. Items to be tested for explosive residue should be placed in a clean metal can and sealed. A factor to not is evidence is time sensitive in the explosion environment and rapidly degrades over a matter of hours. Metal fragments, wires, battery components, unconsumed explosives and related materials are very valuable in reconstructing the device. These items may allow investigators to locate where specific items were purchased and assist in identifying the bomber. All items should be photographed and their location plotted on a crime scene drawing prior to collection. Whenever possible leave identification and collection evidence to post blast investigators.

LAW ENFORCMENT AND SECURITY RESPONSE TO EXPLOSIONS

- REMEMBER TIME, DISTANCE, AND SHIELDING!
- Remember, an explosion may be a distraction while another crime occurs.
- Proceed cautiously, stop outside the damage area and observe.
- Use binoculars when feasible to survey the area.
- Call for the bomb squad early.
- Call for needed mutual aid early into the incident.
- *Enter the blast area with minimal personnel only to save lives!*
- Try to determine who was in the blast area and account for them.
- Stay uphill, upwind, and avoid glass when possible.
- *Think about secondary devices!*
- *Think about ambush with firearms!*
- *Think about suicide bombers in the bystanders or among the injured!*
- Stay out of fogs, liquids, smoke, and mists as they maybe toxic.
- Do not enter an area without proper respiratory protection and protective clothing when multiple persons are down who appear to be having seizures or when a mass illness is present. You may become a casualty.
- Choose the best spot outside the blast area for a command post or triage area, and pick another to avoid preplaced secondary devices or a preplaced ambush.
- Look for suspicious objects, vehicles, bags, briefcases or similar items. Check license plates looking for a rental vehicle, stolen license, stolen vehicle or switched tag. This may be a vehicle bomb.
- Use your public address system to encourage the injured who can walk to come out of the blast area. Send them to the EMS triage area.

- Send minimal personnel in to "load and go" those injured who cannot walk. Expedite their removal from the danger area.
- When feasible have the injured remove their clothing to avoid cross contamination.
- Have the clothing isolated and retained for law enforcement when feasible.
- Leave the dead but keep a list of the injured and the location to which they are transported as their clothing may contain evidence.
- Secure a circular perimeter at least 1.5 times the distance from the center of the blast to the point the damage ends, or at least 300 to 1,500 feet from the point the damage ends.
- Stage Fire and EMS at least 1,500 feet away unless they are needed on scene.
- Broken glass and debris in the street is a good indicator of blast damage.
- Fight fires defensively with remote appliances.
- If no fires or injured persons are present, *do not enter the blast area* until cleared by bomb technicians.
- Summon Urban Search and Rescue Units early if a structural collapse with victims entrapped exists.
- Summon utilities to secure gas, water and electrical power to damaged structures. Accomplish this from outside the blast area when feasible.
- The blast area may contain partially consumed explosives, additional devices, dangerous debris, biohazards contamination and other hazards such that one should enter only to save lives until the hazards are mitigated.
- Treat the blast area as a crime scene with valuable evidence present.
- Keep track of any persons, vehicles or items that are removed from the scene. They may contain valuable evidence.
- Keep entry and exit points open for emergency vehicles.
- Locate and interview witnesses when feasible.

FIRE RESPONSE TO EXPLOSIONS

- REMEMBER TIME, DISTANCE AND SHIELDING!

- Proceed cautiously, stop outside the damage area and observe.
- Coordinate with law enforcement and security.
- Call for needed mutual aid early into the incident.
- *Enter the blast area with minimal personnel only to save lives!*
- Try to determine who was in the blast area and account for them.
- Stay uphill, upwind, and avoid glass when possible.
- *Think about secondary devices!*
- *Think about preplaced ambush with firearms!*
- *Think about suicide bombers among the injured!*
- Stay out of fogs, liquid, smoke, and mists as they maybe toxic.
- Do not enter an area without proper respiratory protection and protective clothing when multiple persons are down whom appear to be having seizures or when a mass illness is present.
- Choose the best spot outside the blast area for a command post or triage area pick another to avoid preplaced secondary devices or preplaced ambush.
- Look for suspicious objects, vehicles, bags, briefcases and similar items and report them to law enforcement.
- Use your public address system to have the injured whom can walk to come out of the blast area.
- Send minimal personnel in using "load and go" EMS tactics for those injured who cannot walk out of the blast area.
- When feasible have the clothing removed of those exiting the blast area to prevent cross contamination. Isolate the clothing.
- Leave the dead but keep a list of the injured and the location to which they are transported as their clothing may contain evidence.
- Stage at least 1,500 feet away from the blast area unless needed on scene to fight fires.
- Fight fires defensively with remote appliances.
- If no fires are present or injured persons down, do not enter the blast area until cleared by bomb technicians.
- Summon Urban Search and Rescue Units early if a structural collapse with victims entrapped exists or is suspected.
- Summon utilities to secure gas, water, and electrical power to damaged structures. Accomplish this from outside the blast area when feasible.
- The blast area may contain partially consumed explosives, additional devices, dangerous debris, and one should enter only to save lives.

- Treat the blast area as a crime scene with valuable evidence present.
- Keep track of any persons, vehicles, or items that are removed from the scene as they may contain valuable evidence.
- Keep entry and exit points open for emergency vehicles.

EMS RESPONSE TO EXPLOSIONS

- Coordinate with law enforcement and security.
- REMEMBER TIME, DISTANCE AND SHIELDING!
- Proceed cautiously, stop outside the damaged area and observe.
- Use binoculars when feasible.
- Call for needed mutual aid early into the incident.
- Notify medical control and activate the mass casualty plan.
- *Enter the blast area with minimal personnel only to save lives!*
- Try to determine who was in the blast area and account for them.
- Stay uphill, upwind, and avoid glass when possible.
- *Think about secondary devices!*
- *Think about preplaced ambushes with firearms!*
- *Think about suicide bombers among the injured!*
- Stay out of fogs, liquid, smoke, and mists as they may be toxic.
- Do not enter an area without proper respiratory protection and chemical protective clothing when multiple persons down who appear to be having seizures or when a mass illness is present.
- Choose the best spot outside the blast area for a triage area and pick another to avoid preplaced secondary devices or ambushes.
- Look for suspicious objects, vehicles, bags, briefcases and notify law enforcement.
- Use your public address system to have the injured whom can walk to come out of the blast area.
- Send minimal EMS personnel in to "load and go" those injured who cannot walk.
- When feasible have the clothing removed of those exiting the blast area to prevent cross contamination. Isolate the clothing.
- Leave the dead but keep a list of the injured and the location to which they are transported as their clothing may contain evidence.

- Follow the appropriate local EMS Treatment Protocol.
- Remember that persons with crush injuries may require large amounts of IV fluids. Contact Medical Control for instructions regarding crush injuries.
- Remember the injured persons clothing may contain evidence.
- Stage at least 1,000 feet away unless needed on scene for injured persons.
- If no injured persons are in need of rescue, do not enter the blast area.
- The blast area may contain partially consumed explosives, additional devices, dangerous debris; you should enter only to save lives.
- Treat the blast area as a crime scene with valuable evidence present.
- Keep track of any persons or items that are removed from the scene.
- Keep entry and exit points open for emergency vehicles.

Chapter 17

BIOLOGICAL, CHEMICAL, AND RADIOLOGICAL WEAPONS

The material for this portion of the text is drawn from data published in the public safety forum by the U.S. Army Chemical School Domestic Preparedness Program. The potential use for weapons of mass destruction (WMD) is a consideration that must be planned for by all law enforcement and security entities. The weapons can be arbitrarily divided into categories of: biological weapons defined as disease-producing bacteria or viruses, and bio-toxic materials such as ricin. Incendiary devices using improvised flammable materials as simple as gasoline, Molotov cocktail or more exotic high temperature accelerants are considered weapons of mass destruction. Nuclear devices will usually appear in the form of a radiological dispersion device or radioactive material dispersed in another manner. Chemical devices dispersing toxic agents such as Sarin or vesicant agents, or improvised chemical agents are a form of weapon of mass destruction. Explosives in the form of bombs, improvised explosive devices (IED) qualify as a weapon of mass destruction.

In many instances, the public safety providers and government officials may not be psychologically prepared to respond to incidents involving WMD. The number of casualties and extent of damage may overwhelm them. Many public safety providers and other officials are not mentally prepared for such incidents. This may prove to be a serious detriment if such an event occurs, as may be a unique experience. Most have not made any mental preparation regarding the mitigation of such incidents. This is of great concern in the planning for WMD incidents. Following WMD incidents, an advisable step is to have critical stress incident debriefing take place for those involved. Such should be conducted by a trained mental health professional.

The probability of employment of some of the agents is low, such as Sarin, however, the employment of some agents, such as bombs, is quite high. This text addresses the probability of such incidents, the nature of groups or persons who may employ such agents, the identification of such incidents, the initial response and mitigation of such incidents.

CHEMICAL, BIOLOGICAL, AND RADIOLOGICAL DEVICE INDICATORS

The conditions below can be important indicators of deployment of a chemical, biological, or radiological device. These indicators should alert the public safety responders to take protective actions.

- Any suspicious incident in which reports of multiple persons are down, complaining of respiratory difficulty, or seizures, think chemical attack. The agent may be a nerve agent or toxic industrial chemical such as chlorine or ammonia.
- Do not enter any area with multiple persons down without respiratory protection and chemical protective clothing.
- Stay out of liquids, mists, fumes, fogs or smoke, as they may be toxic.
- Stay upwind and uphill.
- Consider turning off ventilation systems within structures to prevent agent spread.
- If an item is received that contains a suspect item, isolate the item, segregate any persons possibly contaminated, consider evacuation, and notify the bomb squad.
- Consider any envelope or item containing unexplained powder, Petri dishes, or unexplained liquids, or gas cylinder, to be a chemical or biological weapon.
- Do not perform any decontamination other than washing materials from skin surfaces with soap and water until members of the bomb squad or hazardous materials team is consulted.
- Look for suspicious or out of place gas cylinders, tubing, Petri dishes, dusts, or liquid containers in the vicinity of a ventilation system, return duct or intake duct within structures or along the

exterior of structures.

- EMS personnel should follow the EMS Protocol based solely on signs and symptoms.
- Structural firefighting clothing is not sufficient protection in chemical, biological, or radiological incidents but may serve as sufficient protection for escape from the area.

BIOLOGICAL AGENTS

Biological agents are pathogenic (disease causing) organisms such as bacteria, viruses, rickettsia, or toxins produced by living organisms. They are divided into three groups.

Bacteria and rickettsia are single-celled organisms that produce a variety of diseases in humans, animals, and plants. These agents may be very potent or weak in their effects. Some produce extremely dangerous toxins in the human body causing disability and death. Most respond to antibiotics, however, resistant strains of bacteria are common. Weaponized bacteria may be genetically engineered to be antibiotic resistant and more virulent than natural strains. Some bacteria produce contagious diseases. The latent period between exposure and active symptoms may be several days. This means that one person who is contagious may expose and infect numerous persons with whom they have contact. This is also advantageous to an adversary, as the attack may not be detected for some time.

Viruses are a very small form that uses the host cell as a reproductive site. Only a few antiviral agents are available at this time. These are the most destructive and dangerous of the biological agents.

Many living organisms produce natural toxins. They range from simple toxins such as snake venom to derived toxins such as ricin. Some of these toxins are the most toxic materials known to man. For example, one or two drops of Sarin nerve agent is probably fatal if inhaled or placed on the skin while ricin (derived from castor beans) is estimated to be more than 10,000 times more toxic. A dramatic amount of information exists within alternative literature and on the Internet defining how to derive or manufacture these toxins. Ricin is likely to be used and several incidents have occurred using Ricin. Ricin has been transmitted by mail in a powder form in envelopes.

Dissemination includes a variety of methods, which include but are not limited to aerosol distribution, explosive distribution, food contamination, and skin exposure. The most common method used is by aerosol dispersion to produce an airborne hazard. A simple household, agricultural or industrial sprayer can be utilized. Modern agriculture uses aircraft and land vehicles to dispense pesticides in this manner. The usual size is between two and six microns. This is optimal for atomization and suspension in air. This also is optimal for inhalation of the agent in humans. Units to produce this type aerosol are readily available in the commercial market with no restrictions. These items can be readily stolen from a farm.

Another reasonably simple dissemination method is to contaminate food. Terrorists in Oregon used simple household spray bottles to contaminate salad bars prior to an election such that their candidate would be elected. Salmonella sickened hundreds of persons and the candidate supported by the terrorists was elected. This was accomplished covertly.

Exposures to the skin of injection are two techniques by which biological agents can be introduced. Ricin has been used to assassinate several individuals by injection of amounts less than one-tenth of a gram. Some agents can be directly absorbed via the skin. Another plot uncovered utilized ricin in concert with DMSO, a potent solvent that would have applied the ricin to environmental surfaces and be absorbed through contact. The plot was uncovered prior to implementation and would have failed since the ricin molecule is too large to be carried through the skin. Numerous other methods are available in alternative literature and on the Internet.

Anthrax is another weaponized bacterial agent. It has been a pathogen attacking man and animals for centuries. This is a hardy bacterium that forms an environmentally resistant spore giving it a long life span in its dormant phase. Anthrax is virulent and can cause a large number of deaths from the inhalation form of the disease in those not treated or when treatment is delayed. However, in its natural form anthrax is antibiotic sensitive and reasonably well controlled.

Two common forms of anthrax infections exist, skin and inhalation. The skin form usually is from direct contact and forms a lesion. The person may become septic and die. Untreated it runs a mortality rate of five to 20%.

Inhalation anthrax, which only requires that a few hundred microscopic spores be inhaled, has a one-to six-day incubation period. The initial symptoms are flu-like and are usually mistaken for influenza. The second phase is rapid in onset with respiratory distress and collapse. The mortality rate of untreated cases is between 90 to 95%. Delay of treatment until symptoms appear also has a very high mortality rate. Time from onset of respiratory distress and death is usually less than 24 hours. This agent occurs naturally and has been weaponized.

Plague (yersina pestis), better known as the "Black Death," is an endemic bacteria existing in parts of the United States. This bacterium is flea borne and is transmitted through fleabites. Symptoms include respiratory distress, bloody sputum and fever. It can be rapidly fatal if not treated. The natural form is susceptible to several common antibiotics. This agent occurs naturally and has been weaponized. Tularemia is rarely fatal but can be delivered in aerosol form. It produces influenza-like symptoms and is not usually fatal. Q Fever is a rickettsia that produces influenza-like symptoms. It may produce meningitis (brain and spinal cord inflammation), pericarditis (heart sac inflammation) and myocarditis (heart muscle inflammation). The disease is debilitating but not normally fatal.

Viral agents are smaller and operate differently than bacteria. They force the host cells to reproduce the virus.

Smallpox virus is transmitted via aerosol and is very contagious. The incubation period is seven to 17 days when a pox-like rash begins. It presents influenza-like symptoms. Scabs begin to form on day eight to 14. This agent has been used in biological warfare numerous times including the 1700s when infected blankets were used to decimate a Native American group. This agent has been weaponized.

Venezuelan Equine Encephalitis (VEE) is a mildly contagious disease with an incubation period of one to four days. It presents influenza-like symptoms and is occasionally fatal.

Viral Hemorrhagic Fever (VHF) or hemorrhagic fevers include a variety of deadly diseases such as Ebola, Crimean-Congo Hemorrhagic Fever, Yellow fever, Hantavirus, etc. All these present influenza-like symptoms followed by collapse. The mortality rates ranges from moderate to very high. Some are very contagious but most are only moderately contagious.

Toxic substances can be found from a variety of organisms. Plants, bacteria, snakes, sea life, and fungi produce many toxins.

Botulinum toxins consist of at least seven neurotoxins produced by the bacteria Clostridium botulinum. The classic symptom of this agent is diplopia (double vision). It kills via advancing paralysis and respiratory arrest. Several vaccines are available. This agent has been weaponized. Symptoms usually occur within 24 hours of ingestion. The most common civilian exposure is food spoilage.

Staphylococcal enterotoxins are the cause of food poisoning in improperly prepared food. These toxins cause nausea, vomiting and prostration. They occasionally produce death. It can be dispersed as an aerosol.

Ricin is derived through the caustic reduction from the mash of castor beans. Ricin is extremely toxic and amounts as small as one milligram can cause death. It can be dispersed as an aerosol, liquid or powder. Initial symptoms are influenza-like but rapidly progress to necrotic lesions of the liver, spleen, circulatory collapse, and death. This agent has been used on several occasions and has been found manufactured by domestic terrorist groups.

Mycotoxins (Trichothecene Mycotoxins) are derived or produced by fungi. This agents cause a variety of symptoms such as weight loss, GI bleeding, skin inflammation, hemorrhage, and occasionally death.

Anthrax and Ricin Threats

The most common biological threat will be the threat of anthrax spores or the biological toxin Ricin. Anthrax spores are a hardy form of the bacteria, which if inhaled, ingested or expose to non-intact skin can cause illness. Only a small number of spores are thought to be infective to the victim. Ricin is extremely toxic and small amounts (several milligrams of the pure toxin) can be fatal if inhaled or ingested.

If an item is labeled that it contains anthrax or Ricin with no visible contamination should be placed in a plastic bag and that bag placed in a second bag. The person or persons who have been in contact with the item should immediately wash their hands with soap and water. The area exposed to the item should be isolated and local law enforcement notified. Substantial decontamination should not take place until the situation is evaluated. The bomb squad and local hazardous material unit can offer assistance in these circumstances.

If an item is opened and visible contamination is discovered, the person handling the item should immediately place the item in a location that will prevent the further spread of the material. This may be in a nearby trashcan or on top of a desk. Use any item available to cover the contaminated letter or package to prevent the further spread of contamination. If no visible contamination is present on clothing, the potentially contaminated person should immediately wash their hands with soap and water. Isolate the potentially contaminated person and prevent contact with other persons or areas. The area in which the contaminated item is placed should also be isolated. If a large amount of contaminant is visible, consider shutting down the ventilation system. If visible contamination is on the clothes of the individual, consider removing and leaving the clothing in the area potentially contaminated. The person should shower and wash thoroughly with soap and water including his or her hair.

A variety of items may be used to simulate anthrax spores or Ricin such as talcum, chalk dust, flour or sugar. In most incidents, the powder will be a hoax. Remember, hoaxes are criminal offenses in most jurisdictions. Be familiar with your jurisdiction's policy and handling of these incidents. The primary effort should be directed at controlling contamination spread and confirming that the incident is, in fact, a hoax. Most agencies will have a specific protocol for determining whether on-site testing will be conducted or sending samples to a laboratory for analysis. Decontamination policies may also vary.

A variety of factors may alert responders to the deployment of a biological agent:

1. Threats prior to the incident, verbal, or written. These may be general or specific.
2. Intelligence that indicates a group is biological agent capable.
3. Out of place, unusual equipment such as sprayers, tubing, pressurized gas tanks, Petri dishes, culture or growth media, incubators or similar equipment.
4. Biological Hazard labeling.

Most on-site detection methods are not practical for widespread deployment and are very expensive. However, with the advances in technology and cost decreases, inexpensive and accurate testing is available in the form of field tests which are available. However, many of the on-site tests result in false positive readings.

Bioassay techniques may be required, taking a sample from the living organism to detect some biological agents or their toxins. The federal government and most states have laboratories suitable for this purpose. Many states departments of public health are implementing laboratories for analysis and identification of biological agents.

Mass spectrometry may be needed in some circumstances, usually requiring a laboratory setting for this high technology equipment. This equipment applies electrical and magnetic fields to ions and through their behavior identify the type agent.

Residues from the toxins may be analyzed using gas chromatography in a solvent in a pressurized gas or liquid environment and comparing the behavior of the material with known substances. This may be combined with other techniques and can provide precise data.

- Respiratory protection, usually in the form of at least a rated air purifying respirator, is critical against all dispersed biological agents. The most common terrorist scenario is the dispersal by aerosol to attack via the inhalation route. The minimal respiratory protection is a properly rated air purifying respirator or self-contained breathing apparatus.
- Skin protection consisting of chemical vapor and chemical splash resistant clothing. Protective gloves, preferably nitrile or other rated chemical gloves and footwear are very important.
- EMS personnel should use all available body substance isolation equipment with multiple gloving. EMS providers should be aware that medical latex and vinyl medical gloves are not rated against chemical or biological agents. Nitrile gloves are recommended.
  1. The first priority must be patient decontamination. One of the serious issues that occurred during the Tokyo Sarin attack was cross contamination from patients to first responders and in hospital personnel. Decontamination can be divided into several categories. Many of the injured self referred and off gassed agent in medical settings. Technical decontamination is the decontamination of physical property such as equipment and tools. Self-decontamination is the steps taken by the responder to remove clothing, personal protective equipment, and take protective actions prior to medical evaluation. Expedient decontamination may involve washing with water and soap.
  2. Once gross decontamination is accomplished, emergency

medical treatment can begin. This is one of the few incidents in EMS that patient care is secondary to another procedure. To fail to decontaminate prior to treatment may subject the patient to further contamination and cross contamination of the responder, EMS vehicle, and hospital.

3. Once initial field management is completed, the patient can be transported to a hospital setting. The medical control physician and destination hospitals should be alerted early into the incident to advise them that contaminated patients are being transported to their facility with the nature of the contamination, if known. This allows time for the hospitals to alert staff and prepare decontamination facilities.

4. Definitive care can consist of a variety of treatments, which may include detailed decontamination consisting of treatment with the appropriate prophylactic antibiotic or antiviral agent and other medical treatment.

NUCLEAR OR RADIOLOGICAL DEVICES

A nuclear device might be used for terrorism in several scenarios. The most likely scenario is a Radiological Dispersion Device (RDD) which is a device designed to disperse radioactive materials through conventional means. The usual mode is an explosion since most radioactive materials will tolerate the thermal event of an explosion.

Devices of this nature have been constructed and detonated in other countries. This device usually spreads radioactive material through an explosion. This device may disperse radioactive material over a large area or be limited to a single structure. Alternative devices use a dry form of the radioactive material to contaminate a large area through a non explosive dispersion as in a ventilation system. Such a device was found in Russia in 1995 and was built by a terrorist organization. The device contained approximately 30 pounds of explosives and radioactive Cesium was located in a park.

Effects and dangers of such a device include denial of use of a facility or area, radioactive contamination of persons and property, radiation exposure or burns, injury from the explosive dispersal and long term health effects. The most probable source of radioactive materials will be radiography sources. These are reasonably available materials and theft prone. The same is true for medical isotopes. Neither of these items could contaminate a large area but could produce significant effects over a smaller area such as a structure.

A less likely scenario is the dispersal of radioactive materials through the utilization of sabotage or explosives to cause the failure to containers or containment. Most radioactive materials in sizeable quantities are well contained and their containers are designed to resist explosive attacks. The question of sabotage is somewhat more likely since direct intervention might allow the release of radioactive materials. This is more difficult to prevent and guard against.

A variety of facilities contain nuclear materials. However, typically these facilities have security and are constructed to resist such attacks. Likewise, transport vehicle and containers are similarly constructed. The factor that would facilitate the release such materials would be sabotage or forcible compulsion of personnel to conduct such a release. This is a far more probable scenario than the use of brute explosive force to disseminate radioactive materials.

Examples of the facilities include:

1. Military facilities where nuclear weapons are housed.
2. Nuclear weapons construction and maintenance facilities.
3. Nuclear powered vessels.
4. Nuclear Power Plants.
5. Fuel reprocessing facilities.
6. Nuclear Waste Facilities.
7. Medical facilities using medical isotopes. Medical facilities are located within and nearby counties using medical isotopes.
8. Radiographic sources or facilities using radiographic sources.

The least likely scenario is the detonation of a fission device. These devices in theory could be constructed by a terrorist nation-state and deployed in the United States. This is not highly probable according to most open source intelligence. Even less likely than a theft is the defeating of permissible action links on military nuclear weapons.

This is a very unlikely scenario since the funds and ability to deliver such a device are capable only by a terrorist nation-state. Current open source intelligence indicates this is not a probable scenario. To build a nuclear device a substantial amount of fissionable material is required. The theft and missing quantities of fissionable materials that has been reported are sufficient to construct such devices. There has been seizures of fissionable and weapons grade materials being smuggled from the former Soviet Union. Generally the opinion is that although the amounts of missing materials worldwide are sufficient for construction of such a device, the materials are not in the possession of the same persons or groups with the resources to perform such construction. The minimum amount of uranium to construct a crude nuclear device is approximately 156 pounds. This does not include the shielding and ancillary support devices required to detonate the device. Shielding and the high density make the device difficult to transport with the density by volume of uranium which is more than 18 times that of water. The likely sources of such materials are theft of spent nuclear fuel, theft of fissionable materials, and acquisition through illegal diversion, theft of military weapons, or theft from a storage facility.

The difficulty in construction of the device is far beyond the acquisition of the fissionable material. An exact geometry and sophisticated triggering systems are required. Very specific quantities and configurations of explosives are required. The requirements are beyond any typical terrorist group. However, such resources are available in terrorist nation-states. Several nations that are hostile to the United States do possess nuclear weapons.

The theft of an operational nuclear device is also improbable. Such devices are highly guarded and require sophisticated activation through a permissible action link. More likely is the theft of the fissionable material within the weapon.

The most common threat involving a nuclear device or radiological device will be the threat to utilize such a device. Although radioactive materials are reasonably and easily obtained, there has not been a propensity to deploy these devices. A threat of such an event is more common. If such a threat is received state and federal assistance is mandatory. The resources to conduct high technology clandestine searches for such a device are readily available at the federal level. The lead agency will be the FBI with the Department of Energy's Nuclear Emergency Search Team assisting.

The primary threat for these devices is radiological. Radiological hazards are continuous rather than episodic meaning the threat is an ongoing hazard. The important facet of this hazard is to determine its nature, the specific type of radioactive material or radiation involved and its scope. Radiation detection instruments and assistance from emergency management, state radiological health, and federal agencies may be required. Most bomb squads have rudimentary capabilities of detecting radiation devices or radioactive materials. The Civil Defense type instruments have the crude capability of radiation detection but cannot detect some lower energy medical type isotopes or may read zero in high radiation areas.

The secondary threat is chemical. Most radioactive materials are heavy metals and toxic chemically as well as being radioactive. Medical isotopes and many other commercial radioactive materials are biologically active. This means if absorbed they will concentrate in target organs making them even more dangerous both chemically and for radiological purposes. For instance, cesium is bioaccumulated in the liver.

The most important protection factor is the detection of radioactive materials. The failure to initially detect radioactive materials may result in extensive personnel and equipment contamination. If not detected one cannot take protective measures.

A repeating theme in protection methods is time, distance, and shielding. The **A**s **L**ow **A**s **R**easonably **A**chievable (ALARA) concept was originally conceived to reduce occupational radiation exposure. Components of the concept include reducing the time one is exposed to radioactive material and within radiation areas. With this in mind, spend the shortest amount of time possible in and around contaminated areas or radiation sources.

Another component of ALARA is distance. Stay as far away as is feasible from the radiation source or radioactive materials. Radiation obeys the inverse square law such that doubling the distance reduces the radiation exposure by its square root. Distance is a critical factor in protection from radiation. The North American Emergency Response Guide (NAERG) recommends a minimal isolation area of 25 to 50 meters (80 to 160 feet) in all directions and remain upwind. If a radiation dispersion device with explosive components then the evacuation distance of responders should be extended to a minimum of 1,500 feet.

Shielding at the scene of a radiological emergency can be accomplished using several techniques. The most dangerous radiation at a distance is gamma radiation. Massive physical objects such as structures, vehicles, or similar objects can stop or lessen the strength of gamma rays. The more massive the object the better the protection offered.

Internal contamination occurs when radioactive material is taken into the body. This commonly occurs by inhalation, ingestion, or skin absorption. Another form of shielding is the utilization of personal protective equipment to prevent skin or internal contamination. The utilization of respiratory and skin protection is essential. Responders should take every effort to thoroughly decontaminate themselves and equipment exposed to contamination. Usually washing with soap and water will remove most external contamination. There should be no eating, drinking, chewing or smoking during these incidents. The only positive method to confirm decontamination is by survey with a special type of radiation instrument suited for detection of the specific type of radiation emitted by the radioactive material suspected to be present. Trained radiation monitors and health physics personnel are needed to assure decontamination is complete. Most local emergency management agencies and state level emergency management can provide this resource.

The warning indicators of a nuclear fission device being present are standard radioactive materials warnings, placards, and shielded containers. Older Civil Defense type radiation detection instruments are not capable of detecting shielded fission devices. However, gamma scintillation instruments available from typically local or state emergency management can detect such devices. Sophisticated detection equipment is also available through the Department of Energy Nuclear Emergency Search Team (NEST) or Domestic Emergency Support Teams (DEST). Evacuation and isolation recommendations are addressed in the NAERG.

All explosion sites and any suspicious container should be scanned with radiation detection instruments. Bomb squads carry basic radiation instruments. Older Civil Defense type instruments are usually available but cannot rule out certain types of isotopes. The FBI certified bomb squad should have a gamma scintillation detector, which can be used on an emergency basis. In suspicious circumstances, local or state emergency management should be contacted.

The common concepts of time, distance and shielding should be used. Standard decontamination methods are effective. However, sensitive radiation detection instruments are required to confirm the removal of radioactive material. Important in the protection efforts are respiratory protection and skin protection.

Decontamination must take place prior to anything other than life-saving treatment. Usually standard decontamination techniques will work. However, a sensitive specialized radiation detection instrument is required to detect radioactive contamination. Patients cannot be transported until decontaminated without serious risk to rescuer, EMS vehicle, and emergency department contamination.

Patient management other than lifesaving interventions can proceed once decontamination is complete. Notify medical control and the receiving facility that radioactive contamination has been encountered. This will allow the facilities time to prepare for these patients. There may be a designated facility in your area that is equipped to handle radiation patients or those whom have been contaminated.

Definitive medical care and intensive decontamination must occur in a hospital setting. Critical to this effort is the health physics research to determine the degree and nature of the patient's exposure and contamination and isotopes involved. The state radiological health agency and the hospital's nuclear medicine staff can assist these circumstances.

CHEMICAL AGENTS

Chemical agents use a variety of methods to injure or kill. The most common agent to be employed in a military environment recently and postulated to be the weapon of choice for terrorists is a nerve agent. Sarin, a nerve agent, was used in Tokyo. However, this does not rule out other agents being used. Arbitrary classification used by the U.S. military states that non-persistent agents are present in the environment for less than 24 hours. Persistent agents are classified as remaining in the environment for more than 24 hours. Environmental conditions such as precipitation, wind speed, wind direction, sunlight, humidity and topography dramatically affect agent dispersal and persistence. Persistent agents are usually delivered in a liquid or large droplet form, whereas non-persistent agents are delivered in aerosol form with very small droplets.

Nerve agents are deadly and are some of the more toxic substances known to man. They are extremely toxic in their vapor or liquid states. Small amounts of nerve agents can injure or kill rapidly. These agents attack the acetylcholinesterase enzyme resulting in an excess of acetylcholine and cause death through respiratory arrest. These agents were originally derived from organophosphate insecticides, which might be used as an improvised nerve agent.

Vesicants or blister agents attack any exposed skin, the eyes, and if ingested or inhaled, the gastrointestinal system and lungs. They cause loss of external skin or tissue forming reddened areas and blisters hence the name "blister agent." Their effect can be likened to a very corrosive material in contact with the body part exposed.

Cyanides and blood agents attack the body through disruption of respiration at the cellular level. Death is through respiratory arrest. Industrial chemicals can be substituted for these agents particularly cyanide and cyanide-related compounds. These agents require a relatively high concentration to kill requiring the victims be in the release plume of within a structure.

Pulmonary and choking agents injure and kill by causing pulmonary edema with resultant flooding of the lungs with body fluids. This manner of death is called "dry land drowning." The signs and symptoms are difficult breathing and fluid in the lungs. Common agents include phosgene, ammonia and chlorine. Both the agents and others are readily available in the industrial setting. However, it takes a high concentration to produce death requiring the victims be in the release plume or within a structure. Remember many of these gases are heavier than air in accumulate in low lying areas.

Irritants and riot control agents include tear gas, irritant chemical sprays and other combinations of the agent. Commonly deployed agents are CS tear gas, CS in liquid form, and OC more commonly known as pepper spray. These agents attack the eyes, mucous membranes and lungs. They make physical activity difficult and are not usually fatal; in fact, they usually leave no long-lasting physical effects or injury.

The preferred and most efficient method is dispersal as an aerosol. Aerosols are droplets suspended in air. This allows the suspended agent to be transmitted over a wide area. The device to create the aerosol may be a handheld sprayer, aircraft sprayer, pesticide spray equipment, larger aerosol generators such as vehicle-mounted foggers, or by explosion.

Area contamination for denial of use normally involves a persistent agent in liquid form. This allows the agent to remain in the environment for an extended time contaminating anyone that ventures into the area. The agents more commonly used for this purpose include VX nerve agent, mustard agents, and thickened nerve agents.

COMMON NERVE AGENTS

- GA or Tabun
- GB or Sarin
- GD or Soman
- TGD or thickened Soman
- VX or V agents

The most common route of exposure to nerve agents is through the inhalation of the aerosol form or airborne vapors. Another route is through skin contact with liquid droplets of the agent.

An important aspect of the physical characteristics of the agents is their vapor pressures. This determines how the agent behaves at varying temperatures and whether or not the vapor form of the agent is easily produced.

Vapor pressure refers to the relative amount of vapor that is produced at a certain temperature until equilibrium is reached. The higher the vapor pressure of a substance the more vapor is produced at any particular temperature. Other factors influence the vaporizing of a liquid such as surface area and airflow. The increased surface area, higher airflow and temperature (by heating) will improve the vaporization and formation of aerosols.

The common measurement of vapor pressure is related to standard atmospheric pressure of 760 millimeters of mercury (mm/Hg). Substances that have a vapor pressure of greater that 760 mm/Hg usually are gases at normal temperature. An excellent example is chlorine, which has a vapor pressure of 7,000 mm/Hg. Sarin, a nerve agent, has a vapor pressure of 2.9 mm/hg. Mustard agents usually have vapor pressures of less than one meaning they usually remain in a liquid form.

Some nerve and blister agents are in liquid form and must be vaporized to be properly dispersed. This can be accomplished through use of a mechanical vaporizer to form a vapor or aerosol, or by heating. The Sarin used in Tokyo was not properly dispersed, resulting in minimal casualties. Had it been properly aerosolized, there would have been extensive loss of life.

The most common indicator of nerve agent vapor exposure and inhalation is pain in the eyes and pinpoint pupils (miosis), running nose (rhinorrhea) and difficulty breathing (dyspnea). A large exposure can cause loss of consciousness, paralysis, copious nasal and lung secretions, respiratory arrest, and death.

Liquid nerve agent on the skin presents a different clinical picture. It usually presents with perspiration at the point of exposure, nausea and vomiting, muscle twitching (fasciculation) at the site of exposure, weakness, copious nasal and lung secretions, collapse, respiratory arrest and death. Agents that have a high vapor pressure such as Sarin are more easily decontaminated than VX as VX is an oily liquid. Liquid agents typically have a longer induction period meaning their effects may be delayed whereas those agents inhaled in aerosol form normally have immediate effects.

Warning Signs of Nerve Agent Deployment

These may range from threats or warnings to the observations of responders. Any unusual event involving an explosion with casualties out of proportion to the explosion should be suspect for an explosive dispersal of nerve agents. Any call with a large number of persons having difficulty breathing, down, or having seizures should evoke concern about a nerve agent being used. High value and densely populated structures such as government buildings, shopping malls, stadiums, sports arenas or similar structures with unusual occurrence should raise the responders' index of suspicion. Patients presenting respiratory arrest, with pinpoint pupils (miosis), eye pain, copious nasal secretions (rhinorrhea) and difficulty breathing (dyspnea) should be evaluated for nerve agent exposure. Cross contamination of responders is common in these circumstances mandating the use of respiratory protection and protective chemical clothing.

Detection of nerve agent employment can be the same methods used for the detection of organophosphate pesticides. Actual confirmation of the nature of the nerve agent may require a lengthy process with off-site laboratory analysis.

• M8 or M9 military detection papers.

- Standard hazardous materials detector systems.
- Pesticide ticket detectors.
- Electronic devices.

The most important personal protective measure is respiratory protection. Unless fully protected, the responder should not enter any area in which persons are down, particularly if they are having seizures. The responder should also avoid contact with persons leaving such an area as they may be contaminated and cross contamination following nerve agent exposure is probable. Nerve agents also require complete skin protection. Many responders will have self-contained breathing apparatus available but with proper filters air purifying respirators are suitable for this use (non-confined space). NAERG Guide 153 should be followed when possible. Do not enter any potentially nerve agent contaminated area without appropriate training and protective equipment. Use the protective measures of *time, distance* and *shielding.* Antidotes to nerve agents are Atropine and 2-PAM Chloride in the Mark 1 kit or DuoDote®. These antidotes are available in auto injectors for use by trained responders. EMS personnel with advanced life support capabilities will carry only a small amount of atropine but usually no 2-PAM Chloride. Some EMS providers are carrying self-injecting syringes from these kits. The minimum recommended number of Mark 1 kits or DuoDote® injectors is three per responder. This minimal number would facilitate the self or buddy treatment of a responder exposed to a nerve agent. Remember, those exposed to a nerve agent must be decontaminated prior to treatment and responders must wear protective clothing and respiratory protection. However, the antidote kits can be used without removal of clothing on victims by rescuers whom are properly protected and are exhibiting substantial signs and symptoms such as shortness of breath, rhinorrhea and dimness of vision.

VESICANT OR BLISTER AGENTS

These agents were designed to deny areas and equipment to enemy personnel. They are insidious in that some have little or no initial symptoms followed by serious effects as late as 24 hours after exposure. Examples include:

- Mustard (H)
- Distilled Mustard (HD)
- Nitrogen Mustard (HN1, HN2, HN3)
- Lewisite (L)

The "blistering" effects of mustard agents usually affect the eyes, airway, and skin. Any agent absorbed may affect other body systems such as the gastrointestinal system and liver.

The earliest and mildest form of skin damage is reddening of the skin (erythema). Initial skin sensations may be itching, stinging or burning which resembles a first-degree burn or sunburn. This may appear as early as two hours after exposure or as late as 24 hours after exposure.

The primary damage to the airway will be death (necrosis) of the mucosa or moist parts of the airway. Later damage will include underlying muscle tissue. Respiratory failure is common cause of death in mustard agent exposure.

The eyes are very sensitive to mustard vapor injury. The latent period for the eyes to have symptoms is shorter than the skin. The gastrointestinal tract is very sensitive to mustard agents. The central nervous system effects are not well known. It appears that some of the agents may induce seizures and reports that high concentration exposure may produce seizures. However, this is not a common finding. It appears that exposure to mustard agent may induce lethargic and sluggish behavior as evidenced from data from World War I and Iran. Lewisite is a similar agent to the mustard agents, but its effects are immediate. This agent causes immediate excruciating burning and pain of the eyes and mucous membranes. Later developing symptoms include a similar progress as with mustard agents.

The warning signs may be previous information or threats. Other warning signs may the manifestation of symptoms and signs of mustard agent exposure.

Detection methods can utilize any of the following:

- M8 and M9 paper
- Military detection kits
- Standard hazardous materials detectors
- Electronic devices

Follow the local protocols in dealing with a hazardous materials incident. Make certain that respiratory and skin protection is present. If the specific blister agent is not identified but responders suspect a blister agent present, then follow NAERG Guide 153. Do not enter the area unless you have the appropriate personal protective equipment and training. Remember those contaminated cannot be treated until decontaminated and cross contamination is likely, Use *time, distance* and *shielding.*

CYANIDE AND BLOOD AGENTS

Examples include Hydrogen Cyanide (AC) and Cynaogen Chloride (CK). Exposure to this agent is generally through a vapor or infrequently through a liquid. This agent can be generated without difficulty in an improvised method by adding an acid (usually hydrochloric acid) to a cyanide compound (either potassium or sodium cyanide) or other cyanide salts.

Cyanides react with respiration at the cellular level and prevent hemoglobin from absorbing and transporting oxygen. Concentrations sufficient to kill and injure are quite high requiring the victims be in the gas plume or within a structure.

The symptoms may be collapse; respiratory arrest and occasionally the hypoxia (lack of oxygen) may induce seizures. The treatment is to remove the patient to fresh air and administer the contents of a cyanide antidote kit as directed by protocol. If the patient recovers it will be usually without any long-term effects. Cyanide agents are not usually persistent but clothing of the victims should be removed to obviate any chance of cross contamination.

The only warnings of this attack may be a number of victims down having difficulty breathing. Sneezing is associated with cyanide exposure and initial hyperventilation followed by slowing respiration, collapse, respiratory arrest and death. The difference between this and nerve agent exposure is the miosis (pinpoint pupils), seizures, nasal secretions and lung secretions. These are not seen as often. The detection techniques include military detection kits, standard hazardous materials detectors and electronic meters.

Personal protective equipment including respiratory and skin protection is mandatory. Follow local hazardous materials protocols. If the agent is positively identified as Cynaogen Chloride, then use NAERG Guide 117. If the agent is positively identified as Hydrogen Cyanide, then use NAERG Guide 123. If the agent is not identified then utilize NAERG Guide 123. Several cyanide antidote kits are commercially available.

PULMONARY OR CHOKING AGENTS

Examples include chlorine (Cl_2) which is greenish-yellow gas is commonly commonly used in water purification andmany industrial processes. It is readily available and quite lethal in high concentrations. It was used in World War 1 as a chemical warfare agent with success.

Phosgene (CG) is a gas that has the smell of newly mown hay. It was utilized as a chemical warfare agent in World War 1 with success. Phosgene is lethal in high concentrations and is encountered in fires involving fluorocarbons on occasion.

Another industrial gas which is amenable to use as a chemical agent is ammonia. Ammonia may be found in refrigeration systems, is used as a agricultural fertilizer and as a reagent in the illegal manufacture of methamphetamine. The vapors are corrosive. When labeled as anhydrous ammonia, the gas has minimal water content.

Exposure is through vapors. The chlorine reacts with water in the airway to form hydrochloric acid, which attacks the lungs. Phosgene reacts in a similar manner as does ammonia with forms a strong alkali when water is absorbed. The patient dies from pulmonary edema or fluid in the lungs. This is sometimes referred to as "dry land drowning." The symptoms include airway irritation, dyspnea, chest tightness and pulmonary edema within minutes to hours.

The warning signs may include respiratory distress and collapse of victims. Victims may report a yellowish-green gas, with the smell of "pool chlorine," an ammonia smell or newly mown hay. Detection Methods include military detection kits, standard hazardous materials detection methods and electronic meters. Important protective actions regarding these agents are the use of respiratory protection and skin protection. Victims should be thoroughly decontaminated after removal of their clothing to remove any residual gases or contaminants. Follow standard hazardous materials operating procedures. If the material is identified as chlorine then use NAERG Guide 124. If the material is identified as Phosgene then use NAERG Guide 125. If the agent is unidentified then use Guide 123.

IRRITANT AND RIOT AGENTS

These agents are not commonly considered chemical weapons. They are designed to incapacitate but are infrequently fatal. Extremely high concentrations of the agent or a pre-existing health problem usually exist to cause death. Examples include:

- CS (most common form of Tear Gas)
- CN (Mace)
- OC (pepper spray, oleoresin capsicum)

These agents are commonly known as riot control, lacrimators, pepper gas and tear gas. They usually produce tearing, eyelid spasm, difficulty breathing, burning sensation of the skin and airway, and pain to exposed areas. The agents may be in aerosol form from an evaporative base, may be explosively dispersed as a fine dry powder, or by burning a base substance to produce the dry suspended particles. These agents are not considered persistent in most forms, however persistent forms exist. These agents are readily available in the commercial market, over the counter and by theft of military agents. These agents have been illegally used to deny use of property by introduction into the ventilation system of an area shopping mall. This resulted in evacuation of the mall and several persons were transported to medical facilities. A large aerosol container of OC was used.

The purpose of these agents is stop violent activity with minimal risk to involved persons, incapacitate those exposed rapidly. These agents are very rapid acting with effects seen usually with seconds of exposure. These agents target the mucous membranes. These agents' effects usually rapidly fade upon exposure to fresh air or washing with water.

Detection of these agents includes observation of the classic signs of exposure to an irritant. Other findings usually include the expended containers, which may appear as aerosol cans, gas grenade cartridges, tear gas powder or similar dispensers.

Detection of these agents is based upon empirical data. Laboratory testing after the fact is the usual mode of confirmation of an agent's use. Many of these agents leave a white or yellow powder residue. Many of the OC agents leave a yellow oily liquid stain or residue.

Personal protective measures include respiratory and skin protection. Local hazardous materials protocols should be followed. Decontamination can be accomplished with soap and water and should be accomplished prior to treatment or contact with the victim. One issue to remember is the initial gross symptoms of OC or tear gas mimic exposure to nerve agents.

TACTICS TO MINIMIZE EXPOSURE TO CHEMICAL AGENTS

Follow the same tactic as one would in any chemical, minimizing time spent in the contaminated area. Stay uphill, upwind, and as far as feasible from the incident site. Follow the NAERG Guide recommendations. Shielding consists of the appropriate protective clothing and respiratory protection.

Casualty treatment requires decontamination prior to treatment except to correct life-threatening injuries. Patient management on scene as indicated with efforts toward correcting any air or breathing-related problems. Notify medical control and the destination medical facility of the contamination issue. Transport to the appropriate facility. Definitive care can be rendered in a hospital setting. On scene treatment is symptomatic.

Chapter 18

CLANDESTINE DRUG LABORATORIES

Law enforcement agencies that have not seen clandestine drug laboratories (CDL) will probably face the problem in the near future. The issue is not only the safety of responding officers but also many of these laboratories present significant public safety hazards. Some are discovered following catastrophic fires or explosions.

Clandestine drug laboratories present a variety of hazards, such as persons who may be armed and under the influence of drugs, physical hazards such as an oxygen depleted atmosphere, booby traps, toxic and corrosive chemicals, explosive chemicals, and carcinogens.

The most common exigent circumstance is the unplanned response to a CDL. This may occur when public safety activities reveal a CDL in an unplanned manner. Many CDL are discovered in this fashion. This is an emergency and officers should secure suspects and immediately evacuate the area.

The local clandestine drug laboratory team should be requested and search warrants obtained as needed. The immediate area should be isolated and evacuated with fire and EMS responders alerted and staged out of the danger area. There should be no efforts made to enter the laboratory, tamper with or move any laboratory equipment. Trained clandestine drug laboratory entry team personnel should make the assessment and determine the type and level of personal protective equipment and/or evacuations as needed.

Any suspects taken into custody should have their clothing removed and be decontaminated as determined by members of the clandestine drug laboratory team or hazardous material team. Usually disposable paper coveralls replace their clothing. The usual method is to have the suspect disrobe, including their footwear, don disposable coveralls, and be transported to a law enforcement facility. The suspect is then showered. All potentially contaminated items shall be left on site.

There may be a variety of physical hazards present in a CDL. These hazards can include a variety of hazards that present a threat. The most dangerous and unpredictable hazard is armed persons present. These individuals may be armed and under the influence of drugs. Fire and explosion hazards are significant as volatile solvents are used. Some of the chemicals used are also flammable, reactive or explosive. Incompatible chemicals are usually present and are reactivity hazards.

Booby traps and bombs are found in the laboratories and the approaches to them. Be alert for trip wires, chemical booby traps, booby traps with fishhooks, booby-trapped refrigerators, deadfalls and pits, etc.

Below grade or tightly sealed laboratories may present an explosive, toxic or oxygen depleted atmosphere.

- Other hazards include electrical hazards, which may present a fire, and explosion hazard through sparks.
- Fire and explosions may occur from volatile solvent vapors.
- Compressed gases or ammonia present cryogenic hazards.
- Oxygen displacement from phosphine gas or solvent vapors may occur.
- Chemicals in a CDL are usually toxic. The primary exposure includes inhalation or skin contact. The chemicals may include cyanide, red phosphorus, phosphine, mercuric chloride, chloroform, methanol, hydrogen, ammonia, etc.
- Acids and caustic materials can cause burns. Several acids such as hydrogen chloride, hydrochloric acid and sulfuric acid that can be found in laboratories along with other caustics include lye and sodium hydroxide.
- Hydrogen peroxide and hydrogen gas present reactivity hazards. If combined with the wrong chemicals they can ignite or explode.
- Sodium and lithium will ignite and explode if contacted by water. Many basic materials may explode if mixed with an acid.

LABORATORY TYPES

Varying types of laboratories may present different or special hazards such as those listed below:

Red Phosphorus Laboratory

- Solvents such as acetone or toluene, which are flammable.
- Red phosphorus and phosphine gas is very toxic and displaces oxygen.
- Iodine and hydriodic acid, which are very corrosive and toxic.
- Sodium hydroxide or lye, which is very caustic.
- Sulfuric acid or hydrochloric acid both, which are corrosive.

Look for:

- Matches and solvent cans
- Red stains from iodine
- Lye or sodium hydroxide
- Sulfuric acid and salt
- Hydrochloric acid
- Hypophosphorus acid

Thionyl Chloride Laboratory

- Ephedrine or Sudafed.
- Flammable solvents such as chloroform, methanol or ethanol are used.
- Hydrogen gas is used which is very explosive.

Look for:

- Solvent cans.
- Hydrogen cylinders, usually red in color and is an explosion hazard.
- Look for ephedrine or Sudafed.

Ammonia Laboratory

- Solvents involved such as methanol or ethanol, which are flammable.
- Water reactive metals such as sodium and lithium are involved.

These are water reactive and explosive:

- Ammonia usually in anhydrous ammonia form is used and is toxic, a cryogenic hazard, caustic, and explosive.
- Hydrogen chloride gas or hydrochloric acid is involved. This is a strong acid.

Look for:

- Gas cylinders of ammonia.
- Solvent containers.
- Lithium batteries or gray colored metal.
- Look for cylinders of hydrogen chloride.
- Look for gas generators using rock salt and sulfuric acid.

Phenyl-2-Propanone (P2P) Laboratory

- These laboratories use very toxic and corrosive reagents such as mercuric chloride and methylamine.
- Flammable solvents such as alcohol are used.
- Hydrogen chloride gas or hydrochloric acid is used.

Look for:

- Solvents.
- Methylamine or mercuric chloride.
- Aluminum foil.
- Hydrogen chloride cylinder or hydrochloric acid.

Phenylacetic Laboratory

- Uses phenylacetic acid, which is mixed with acetic anhydride.
- Caustic sodium hydroxide or lye is used in this process.

Look for:

- Phenylacetic acid or acetic anhydride.
- Lye or sodium hydroxide.

Methylenedioxyamphetamine (MDA) or Ecstasy Laboratory

- Uses isosafrole as a precursor.
- Also used are formic acid and hydrogen peroxide.
- Sulfuric acid and methanol can be used.
- Benzene, which is a suspected carcinogen, may be used.

Look for:

- Isosafrole.
- Hydrogen peroxide.
- Benzene.
- Ammonium formate.

Phencyclidine (PCP) Laboratory

- Piperidine is a precursor chemical.
- Cyanide is used which is extremely toxic in the atmosphere or, by contact.
- Sodium bisulfate, magnesium that is flammable, and iodine, which is toxic, is also used.
- Petroleum ether may also be used.

Look for:

- The listed chemicals.
- Petroleum ether.

Gama Hydroxy Butyrate (GHB) Laboratory

- Gama butyrol lactone is a precursor chemical.
- Lye or sodium hydroxide may be used.

Look for:

- Gama butyrol lactone.
- Lye or sodium hydroxide.

ANHYDROUS AMMONIA

Anhydrous ammonia is a common agricultural product and presents some significant hazards when associated with clandestine drug laboratories. Anhydrous ammonia is stolen and used in clandestine drug laboratories. Ammonia is stored and shipped in containers ranging from a few pounds to large pressurized tanks. It may be carried by trucks, rail car or stored in large fixed tanks down to small agricultural tanks. Occasionally anhydrous ammonia is used in large refrigeration systems and in the manufacture of fertilizer.

The theft and diversion of anhydrous ammonia is an issue that law enforcement and security officers encounter. Anhydrous ammonia is used as a chemical for the manufacture of methamphetamine. Anhydrous ammonia presents a significant hazard in several respects. Officers may find during a theft that valves may be left open or broken off. This can result in large amounts of the anhydrous ammonia being released. The persons injured in these thefts may seek medical attention. Reports of persons with cryogenic (freezing) injuries or lung injuries under suspect circumstances should be investigated. These injuries may provide a link to a clandestine drug laboratory.

Anhydrous ammonia is classified by the U.S. Department of Transportation as a non-flammable gas Class 2.2. The normal placard is a green background placard with the drawing of a compressed gas cylinder. The four digit code is "1005" which is sometimes on the placard. Tanks may have "Anhydrous Ammonia" printed on them. Although labeled as non-flammable, anhydrous ammonia is flammable and explosive under special circumstances. When confined and exposed to a high temperature ignition source, it can explode. If stored in a compressed gas cylinder or tank it presents a cryogenic (freezing) hazard. Anhydrous ammonia is both corrosive and toxic. If released from a pressurized cylinder anhydrous ammonia can produce temperatures as low as minus 28 degrees Fahrenheit. When released from a compressed state anhydrous ammonia expands about 850 times. This product is manufactured with no water present. As it enters the environment, it concentrates in areas with water. The chemical produced when it combines with water is corrosive and toxic ammonium hydroxide. This gas is colorless but has an easily recognized warning odor. Breathing the fumes allows the toxic ammonium hydroxide to be produced in the lungs causing significant injury from the strong alkali properties. The eyes are also a target for the gas. Serious eye injuries can result from exposure to this agent.

Anhydrous ammonia is stored in a variety of containers when manufactured, however, when stolen thieves also use both common and unusual containers. Other pressurized tanks such as refrigerant, LPG tanks or other pressurized tanks may be used. Some of these tanks have a non-sparking brass valve system, which is attacked chemically by the anhydrous ammonia. The chemical reaction leaves a blue-green residue. This reaction if allowed to proceed will eventually cause the valve system to fail resulting in the release of the anhydrous ammonia. Other containers may include thermos-type containers and coolers. Use caution if an odor of ammonia is detected.

If anhydrous ammonia is encountered, law enforcement or security officers should use great care. Usually there will be a smell of ammonia present. This should provide a strong warning. An immediate evacuation of the area is recommended. Consult with the clandestine drug laboratory team or the hazardous materials team. Do not handle or move the container. Stay uphill and upwind of the site if operationally feasible and evacuations should be based upon the amount of anhydrous ammonia believed to be present, the surrounding terrain and occupancy. The businesses that use and store anhydrous ammonia can be identified through your local emergency management office and being familiar with their locations is important should an incident involve a release.

If exposed to anhydrous ammonia immediately evacuate the area. The eyes and lungs are the target organs attacked due to their high water content. If discomfort is noted in the eyes obtain medical attention. Field decontamination can be accomplished by flushing the eyes with water for at least 20 minutes. Removal of clothing that may contain the vapor or ammonium hydroxide is recommended. Exposure to the liquid form of anhydrous ammonia may result in cold injury. If clothing is frozen to the skin, flush with water which should allow the removal of the clothing. Leave injured areas exposed to open air. This allows vapor to off gas from the injured area. Those exposed to vapors with coughing or respiratory discomfort should be transported to a hospital. Field treatment should include high flow oxygen. Skin contacts in the facial areas should be flushed for at least 20 minutes with water. Clothing needs to be removed and decontamination in a shower may be needed. If exposure to liquid anhydrous ammonia has occurred, thermal injury due to freezing may have occurred. Clothing should then be removed and the injured areas left exposed to air to enhance ammonia vapor dispersion from tissue. Those exposed to the vapors that have respiratory complaints should be administered high flow oxygen and transported to a medical facility. Other signs of anhydrous ammonia beside the strong warning smell include brown vegetation and dead animals such as birds.

CLANDESTINE LABORATORY BOOBY TRAPS

The type booby traps found at these labs range from buried boards with nails through them to sophisticated chemical devices. Passive traps include: pits containing buried boards with nails, fish hooks strung on monofilament line at eye level, and rugs covering pits, animal traps, knives protruding from walls, pits filled with poisonous snakes. Attack trained dogs are commonly used for alerting purposes and for defense. In rare circumstances poisonous snakes, including exotic imported snakes and alligators have been used to defend laboratories. Active mechanical devices include trip wires with a "loaded" device to spring sharp objects into the intruder and devices to drop sharp or heavy objects on the intruder.

Explosive devices may include pressure activated explosive mines, trip wire operated firearms, pipe bombs activated by opening doors. Explosives concealed in VCR tapes activated by placing the tape in the VCR, improvised land mines, light bulbs filled with powder or gasoline, explosives wired to a refrigerator light bulb, and explosives wired to light bulbs, switches or contained within flashlights. Incendiary devices may include electrically activated jugs of gasoline, light bulbs filled with gunpowder or gasoline, and self-igniting Molotov cocktails. Chemical booby traps can include trip wire activated cyanide gas producing traps where an acid combines with a cyanide compound, and traps to release flammable chemicals or mix self-igniting mixtures.

Armstrong's Reagent, a mixture of potassium chlorate and red phosphorus, is found in crumpled balls of aluminum foil. Movement, friction, shock, sparks or heat will detonate them with small marble size taking off fingers and larger sizes causing serious injury. Concealed firearms in the laboratory are common. Trails or common approaches can be "mined." Extensive fortifications and fences are frequently encountered. Acoustical alerting devices including vibration sensors, passive infrared sensors, low light vision enhancing equipment, and similar devices may be in use at more sophisticated laboratories. Some are found following complaints of strange odors ("chemical smell," such as ether, MEK (fingernail polish remover), toluene, acetone or a urine odor). Look for unusual vehicular or foot traffic, chemical containers, fortifications, strong fences with no livestock, surveillance equipment at or near entrances or access points, intrusion detecting devices, persons patrolling the property, and residents evasive about occupation.

Chapter 19

LOGISTICAL SUPPORT OF BOMB SQUADS

Bomb squads may require the support of public safety personnel for a variety of resources. Usually if a bomb pre-plan has been conducted with facility security, they may be familiar with their needs. Early notification of the bomb squad commander of incident to inquire what, if any, logistical assistance can be provided is important.

Some typical requests from the facility may include:

- Security isolation and evacuation of the affected areas.
- Coordination with facility management.
- Building plans and drawings.
- Access to facility physical plant and maintenance personnel.
- The location of hazardous materials, sensitive equipment, natural gas and utility connections.
- A location to establish a command post with a landline telephone.
- Ice and potable water.
- Assistance in moving bomb squad equipment.

Ice may be used for cooling vests if bomb technicians have to use a bomb suit. The potable water is used to maintain hydration. Building plans allow bomb squad members to determine what areas require evacuation and the damage potential should a device detonate. The plans also allow the determination of a route of removal should the device need to be moved and route of access for a bomb-suited technician. A landline telephone can be used to secure a connection for computer or voice communications. Maintenance and engineering personnel can assist in the reading and interpretation of building plans and provide insight into the location and surrounding areas near a suspect device.

Fire and EMS personnel may be asked to stage at a nearby location or requested to come on-site to directly assist in support operations. The requests for logistical support may include:

- Preparation for response to provide fire suppression and rescue should a high order detonation occur.
- Treatment of various injuries sustained during the course of operations.
- Insertion of an intravenous catheter (heparin or saline lock) in the bomb technician prior to the donning of the bomb suit.
- Treatment of dehydration or heat injury following activities in a bomb suit.
- Assistance in donning and doffing bomb suit, chemical suit, respirators or self-contained breathing apparatus.
- Assistance in decontamination if a chemical, biological or radiological device is encountered.
- Hazardous materials team support.
- Assistance in establishing a "grab" team to rescue a downed bomb squad member in a bomb suit.

Typical requests will be within the mission of the fire or EMS agency such as fire suppression activities, rescue or rendering medical assistance. An explosion environment may complicate these roles. This will be a dust laden, debris filled, and dark environment if inside a structure. Normal landmarks such as walls and hallways may be gone with structural collapse. Any open wounds will be clogged with dust and many victims may have eye injuries due to the dusty environment. Those involved in the explosion may have hearing loss and numerous open wounds from missile injuries. They may be dazed and have head injuries from blast overpressure or impact. Overpressure may cause injuries to hollow organs such as the sinuses, lungs and intestines. Victims must be in close proximity to a bomb to receive overpressure injuries. Shrapnel injuries can be received at substantial distances and may appear as superficial surface injuries but can overlie a more serious deep organ injury. All penetrating injuries should have hospital evaluation.

Assistance with donning and doffing of bomb suits, chemical suits, respirators and self-contained breathing apparatus is somewhat familiar to fire service personnel. Bomb squad members can show fire service and EMS personnel what assistance they may require in suiting up a bomb squad member or removing the bomb suit.

Some bomb squads have a medical protocol that requires the bomb technician to have an intravenous catheter usually a saline or heparin lock catheter placed (cephalic or basilic vein) on the ventral surface of the forearm. Some bomb technicians prefer to use the dorsum of the hand for an intravenous site so they can be rehydrated without removing the bomb suit. This particular site impairs mobility and typically an excellent idea is to remove the bomb suit to enhance cooling. The forearm location minimally interferes with mobility and manual dexterity and has large bore veins. This maneuver is to facilitate the rapid intravenous replacement of fluids lost due to perspiration while the bomb squad member is in the bomb suit. Bomb suits typically weigh from 75 to 100 pounds. Endurance times depend upon the individual, the environmental conditions, and exertion needed to perform tasks. Usual limits of bomb suit stay times range from 15 to 30 minutes. During this time a significant amount of body fluids are lost. It may take one to two liters of intravenous normal saline to replace and rehydrate the bomb technician. Some bomb squads actually pre-load the member with as much as a liter of intravenous fluid prior to donning the bomb suit to enhance endurance and shorten recovery time. Consult with medical control prior to undertaking any invasive procedures.

If there appears to be a hazardous material, chemical, radiological, or biological weapon involved, the bomb squad may request assistance in containment and identification of the substance involved from the fire department hazardous materials team. All bomb technicians who graduated from the Hazardous Devices School operated by the Federal Bureau of Investigation at Redstone Arsenal have training regarding weapons of mass destruction and their mitigation.

One of the more difficult aspects of bomb technician operations is the removal of the bomb technician whom is down due to an explosion or from medical reasons. One will find it usually is not feasible due to the hazardous environment to remove the bomb suit on site. It can be necessary to remove the bomb technician while wearing the bomb suit. This is a bulky and heavy garment. If the bomb technician is not breathing or the ventilation blower is not working on the bomb suit, expedient methods will need to be used to remove the bomb suit helmet. If the incident involved a fall or explosion, spinal precautions will need to be taken.

A team typically of six personnel will be needed to load the bomb technician onto a spine board or other stretcher. Some agencies use a flexible stretcher such that rounding corners or in tight confines, removal is expedited. If a flexible stretcher is used, the rescuers are depending upon the bomb suit itself to provide spinal immobilization. The risk of aggravation of spinal injuries must be weighed against the immediate hazards presented by the environment.

Once out of a hazardous environment removal of the bomb suit taking spinal precautions as necessary is essential. Listed below is a sample technique for removal of the MED-ENG SRS 5 bomb suit, which is commonly in use by FBI certified bomb squads throughout the United States. Removal of the bomb suit requires practice and may be baffling to unfamiliar personnel. *Bomb suits cannot be cut off. They must be systematically removed!* This is the reason that pre-planning with bomb squads by fire and EMS providers is essential.

PROCEDURE FOR EMERGENCY REMOVAL OF BOMB SUIT

SRS 5 Bomb Suit (see Figure 19-1)

Figure 19-1. The SRS 5 bomb suit worn over a chemical suit and with an air-purifying respirator by a bomb technician responding to a suspected chemical device.

- Log roll onto back if not already on their back.
- Use in-line stabilization of the head, neck and spine throughout the procedure.
- If spinal or neck injury is suspected, maintain the head in the neutral position.
- Loosen the lower helmet knobs by turning and remove top knobs by turning while keeping head and neck in line with the body.
- Remove the helmet face shield carefully by lifting upward.
- Maintain the head and neck in line by grasping under the jaws and remove the chinstrap by unthreading the buckle. Remove the helmet upward.
- Gently remove the helmet while maintaining the head and neck in line.
- Pull the back T-straps, one on each side.
- Pull the front T-strap.
- Split the suit open and place the front portion to the bomb technician's right.
- Unzip the arms.
- Remove the diaper. Find the double Velcro straps at the belt line and single straps on the thigh. Fold the diaper toward the feet.
- Unfasten both shoulder straps.
- Remove the waist strap, a double Velcro strap.
- Unzip the legs and remove the single straps on the lower leg.
- Log roll the bomb technician onto the spine board.
- Apply a cervical immobilization device.
- Secure to the spine board appropriately.

BOMB SQUAD RESOURCES

Law enforcement and security officers face a variety of unusual situations that mandate innovative and unique solutions. The utilization of resources in these circumstances requires the law enforcement or security officer to have a comprehensive knowledge of the available resources and their capabilities.

A unique resource can be a FBI certified bomb squad. The squads have highly trained and well-equipped bomb technicians. Bomb squads may be able to assist in non-traditional roles. They can assist in identifying and formulating a plan for removal of explosive chemicals such as those found in school chemistry laboratories. Many squads have explosion resistant bomb vessels so that dangerous items can be transported to a safe disposal area. Including the bomb squad as part of a multidisciplinary safety assessment team of school, government structures, and private industry will assist in bomb threat pre-plans.

Bomb squads usually are equipped with two or more instruments that can detect explosive gases, volatile organic compounds, and other toxic gases. Bomb squad members are trained to operate the instruments and are familiar with operating in toxic environments.

Many bomb squads provide intelligence and threat assessment for law enforcement and can be of assistance in developing a threat picture. Occasionally, bomb squads have bomb technicians trained extensively in post blast investigation and these investigators can provide valuable technical assistance in the investigation of non-criminal explosions. Bomb technicians are trained in response to and the mitigation of weapons of mass destruction from explosives. Most have had additional training and are familiar with chemical, biological and radiological weapons. These personnel can serve as a resource in planning for and responding to incidents of this nature.

Bomb squads can perform portable x-rays and many have fiber optic surveillance systems. Some can perform explosive breaching. Identification of your local and regional bomb squad and their resources is essential, not only for their traditional role in bomb threats, bombings, and found explosives, but also for the unique instrumentation and other skills available.

Appendix

TELEPHONE RESOURCE LIST

This section should be completed and then reviewed annually for accuracy.

Local Law Enforcement:_____

Bomb Squad: _____

Military Explosive Ordinance Detachment:_____

Telephone Company Security Unit/Trace Center:_____

FBI:_____

ATF: _____

Fire:_____

Hazardous Materials Unit:_____

EMS:_____

Fire Marshal: _____

Local Emergency Management: _____

State Emergency Management:_____

Local Public Health: _____

State Public Health:_____

State Police: _____

State Radiological Health: _____

Domestic Preparedness Office: 800 368-6498 or (202) 324-6928

State Environmental Protection: _____

Federal EPA: (202) 260-2090

Military Police:_____

United States Postal Inspectors:_____

Local Postmaster:_____

Local Package Delivery Services:_____

Mortuary Services: _____

Critical Incident Debriefing:_____

GLOSSARY OF COMMON BOMB OR INCENDIARY COMPONENTS TERMS

A wide variety of materials and chemicals can be used to manufacture bombs, explosives and incendiaries. Some of the more common items are contained below. Many of these ingredients may also be found in clandestine drug laboratories where they are used to manufacture illicit drugs.

ACETONE–Used in making explosives, dissolves many explosives, HMTD, and triacetone triperoxide (TATP).

ACIDS–Used in the making of explosives, incendiaries and chemical devices.

ALCOHOL–Used in manufacturing explosives and mercury fulminate or within bottle bombs when mixed with HTH.

ALUMINUM FOIL–Used in bombs, incendiary, or chemical devices.

ALUMINUM POWDER–Can be used in explosives and incendiary mixtures.

AMMONIA–Used in making ammonium nitrate, ammonium triiodide, and TACC.

AMMONIUM NITRATE–If sensitized can be made explosive if marked as an oxidizer. Common forms are AN, 34-0-0 or 3400 fertilizer.

AMMONIUM PERCHLORATE–An ingredient used in high temperature accelerants.

AMMONIUM TRIIODIDE–Ammonia and iodine crystal that is a primary explosive which is very impact sensitive.

ANFO–An ammonium nitrate fuel oil mixture containing ammonium nitrate 94% with 6% fuel oil and is classified as a blasting agent.

ANHYDROUS AMMONIA–used commonly in the manufacture of methamphetamine

BALL BEARINGS–Can be used in shrapnel or in a trembler switch initiator.

BARIUM NITRATE–Can be used in explosives and incendiaries as an oxidizer.

BARIUM PEROXIDE–Can be used in explosives and incendiaries as an oxidizer.

BATTERIES–Used to power electronics or detonate blasting caps.

BALL BEARINGS–Used for shrapnel and can be contained or attached externally to the bomb.

BENZENE–Can be used in making explosives.

BLACK POWDER–A low explosive consisting of potassium nitrate, sodium nitrate, sulfur and charcoal mixture.

BLASTING CAPS–Used as a detonator in bombs or incendiaries and can be electric, burning fuse type or non-electric.

BRAKE FLUID–Can be used in incendiary devices as a fuel.

BULBS (light or flashbulbs)–Can be used as an igniter when connected to battery or other electrical source.

CHARCOAL–Used in making gunpowder or as a sensitizer of ammonium nitrate.

CHLORATES–Can be used as a component of explosives or incendiary devices as they are powerful oxidizer.

CHLORINE GAS–Can be used in making explosives or as a chemical agent.

CITRIC ACID–Can be used in making triacetone triperoxide a primary explosive or HMTD.

CLOTHESPINS–Can be used in booby trap.

CO₂ CARTRIDGE–Can be used a low explosive bomb container.

CONDUCTIVE RUBBER KEYPADS–These devices make excellent pressure pads for booby traps.

COPPER SULFATE–Can be used in making TACC.

COPPER TUBING–Can be used as a bomb container.

CYANIDE–Can be used in chemical devices.

CYCLOTRIMETHYLENETRINITRAMINE–A secondary explosive known more commonly as RDX.

DIESEL FUEL–Can be used as a sensitizer for ammonium nitrate to create ANFO blasting agent.

DOWELS–Usually wooden, are used for compressing explosives in pipe bombs.

ELECTRONIC COMPONENTS–Can be used as remote initiator such as radio remote control for model airplanes.

FERTILIZER–Can be used as an explosive if sensitized ammonium nitrate 34-0-0 grade.

FLARES–Can be used as an incendiary device component.

FUSE–Professional or homemade for ignition, source for bombs or incendiaries is a common use.

GASOLINE–used as fuel for Molotov Cocktails or other incendiary devices.

GLASS CONTAINERS–used for incendiary devices or bottle bombs.

GLYCERIN–Normally used as incendiary and making nitroglycerin. It ignites on contact with potassium permanganate making it useful as an incendiary component.

HTH–High Test Hypochlorite can be used as an incendiary when mixed with brake fluid. This is a self-igniting hypergolic mixture.

HEXAMINE–Is used in manufacturing the explosive HMTD.

HEXAMETHYLENETRIPEROXIDE–explosive made from hexamine, sulfuric acid and hydrogen peroxide to produce HMTD.

HMTD–See above.

HYDROCHLORIC ACID–Can be used in chemical devices and bottle bombs.

HYDROGEN PEROXIDE–Normally used in the manufacturing of triacetone triperoxide and is a very unstable explosive.

ICE–Used for cooling during making explosives.

IODINE–Used in making explosives such as ammonium triiodide.

IRON OXIDE–Used in thermite as a powerful incendiary.

LEAD ACETATE–A substance used in making lead azide a primary explosive.

LEAD AZIDE–Lead acetate and sodium azide mixed from this primary explosive.

MAGNESIUM–Used in some explosives and incendiaries.

MATCH HEADS–Used in incendiary devices as a fuel.

MATCH STRIKER STRIPS–used to obtain red phosphorus in the manufacture of methamphetamine or incendiary devices.

METAL DUSTS–Used in sensitizing explosives and incendiary devices.

MERCURY–Used for mercury switches and used in manufacturing the primary explosive mercury fulminate.

MERCURY FULMINATE–Made by mixing ethyl alcohol, sulfuric acid, nitric acid and mercury to form this primary explosive.

MERCURY SWITCH–Used anti-tamper device on some bombs.

METAL TUBES–These small diameter tubes can be used for making homemade blasting caps.

MICRO SWITCH–This device is useful in booby traps.

MINERAL OIL–Can be useful in the manufacturing of explosives.

MONO FILAMENT LINE–Normally used for booby trap trip wires.

MOTAR AND PESTLE–Normally is used to mix chemicals and incendiary mixtures.

NAILS–Can be used for shrapnel, particularly hard masonry nails.

NITRATES–Many nitrates can be used as explosives, oxidizers or in incendiaries.

NITRIC ACID–Useful in making explosives and chemical devices and one of the more common acids used for this purpose.

NITROGLYERIN–Made from nitric acid, sulfuric acid, and glycerin while using an ice bath to cool the reaction.

NITROCELLULOSE–This is a powerful incendiary material that can be used in explosives by mixing cotton, nitric acid, and sulfuric acid.

NITROGEN TRICHLORIDE–Mixing ammonium nitrate, hydrochloric acid, and potassium permanganate produces this explosive.

NITROMETHANE–A sensitizer for ammonium nitrate and creates a blasting cap sensitive secondary explosive.

NITROPROPANE–A sensitizer for ammonium nitrate and creates a blasting cap sensitive secondary explosive.

OIL–Used as sensitizer for ammonium nitrate and the manufacturing of explosives or incendiary devices.

PIPES–Used for bomb containers. May be PVC or metallic including copper tubing.

PERCHLORATES–These are strong oxidizers that can be used in the manufacture of explosives or incendiary devices.

PERSON BORNE IMPROVISED EXPLOSIVE DEVICE (PBIED)–A suicide bomb usually concealed upon the person or within an object carried by a person.

PETROLEUM PRODUCTS–Can be used as incendiary or sensitizer for ammonium nitrate.

PHENOL–Can be used in making explosives.

PHOTOCELLS–May be used in booby traps.

PICRIC ACID–An explosive made from phenol, sulfuric acid, and nitric acid.

POTASSIUM CHLORATE–Can be used in gunpowder, incendiaries, and is explosive when mixed with petroleum jelly.

POTASSIUM NITRATE–Can be used in making explosives and incendiaries.

POTASSIUM PERMANGANATE–A strong oxidizer used in making explosives and incendiaries. It ignites on contact with glycerin.

PROPANE–Can be used as a fuel air explosive when dispersed by explosives.

RADIO CONTROL DEVICES–Can be used as a remote detonator. RDX-Hexamine mixed with nitric acid in an ice bath can make this secondary explosive. See cyclotrimethylenetrinitramine.

REED SWITCH–Can be used as a part of a booby trap

ROCKET MOTOR IGNITERS–Model rocket igniters can be used in bombs or incendiaries as igniters.

RED PHOSPHORUS–Can be used as an incendiary and explosive when mixed with potassium chlorate. Has a blood-red appearance and called Armstrong's Reagent.

SMOKELESS GUNPOWDER–Used as an explosive particularly in pipe bombs.

SOAP–Can be used as gasoline thickener in incendiary devices.

SODIUM AZIDE–Used in making lead azide a primary explosive.

SODIUM CHLORATE–A strong oxidizer used in making TACC and other explosives.

SODIUM NITRATE–A strong oxidizer used in manufacturing explosives.

SODIUM PEROXIDE–This is a strong oxidizer that ignites on contact with sugar and is useful in making incendiaries.

STRAWS–Can be used as a fuse when filled with gunpowder.

STRONTIUM NITRATE–This is a powerful oxidizer that may be used in manufacturing explosives.

SUGAR–Can be used in explosives or incendiary devices. Ignites on contact with sulfuric acid if mixed with a strong oxidizer such as potassium chlorate.

SULFUR–This is used to make black powder.

SULFURIC ACID–Can be used as an initiator for incendiary devices and for the manufacture of explosives, HMTD, and triacetone triperoxide.

TA CC–This is an improvised explosive made from sodium chlorate, copper sulfate, liquid ammonia, and alcohol.

TAPE–Used to secure components and wrap a device.

TATP–Triacetone triperoxide, a primary explosive made from citric acid, hydrogen peroxide, and sulfuric acid and is very unstable and flame sensitive.

THERMITE–Aluminum, iron oxide mixture with barium peroxide as an oxidizer that produces a powerful incendiary that literally produces molten iron

TIMING DEVICES–Can be a mechanical clock or electronic timer.

TRIACETONE TRIPEROXIDE (TAPP)–This is a primary explosive.

VCR–Can be used as a timer for a bomb or incendiary device.

WIRE–Used for electrical components.

INDEX

175